Pain and Emotion in Modern

Palgrave Studies in the History of Emotions

Series editors:

David Lemmings, Professor of History, University of Adelaide, Australia

William M. Reddy, William T. Laprade Professor of History, Duke University, USA

Palgrave Studies in the History of Emotions includes work that redefines past definitions of emotions; reconceptualizes theories of emotional 'development' through history; undertakes research into the genesis and effects of mass emotions; and employs a variety of humanities disciplines and methodologies. In this way it produces a new interdisciplinary history of the emotions in Europe between 1100 and 2000.

Pain and Emotion in Modern History

Edited by

Rob Boddice
Department of History and Cultural Studies,
Freie Universität Berlin, Germany

palgrave
macmillan

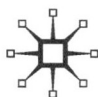

First published 2014 by
PALGRAVE MACMILLAN

Palgrave Macmillan in the UK is an imprint of Macmillan Publishers Limited,
registered in England, company number 785998, of Houndmills, Basingstoke,
Hampshire RG21 6XS.

Palgrave Macmillan in the US is a division of St Martin's Press LLC,
175 Fifth Avenue, New York, NY 10010.

Palgrave Macmillan is the global academic imprint of the above companies
and has companies and representatives throughout the world.

Palgrave® and Macmillan® are registered trademarks in the United States,
the United Kingdom, Europe and other countries.

ISBN 978-1-349-47613-8   ISBN 978-1-137-37243-7 (eBook)
DOI 10.1057/9781137372437

This book is printed on paper suitable for recycling and made from fully
managed and sustained forest sources. Logging, pulping and manufacturing
processes are expected to conform to the environmental regulations of the
country of origin.

A catalogue record for this book is available from the British Library.

Library of Congress Cataloging-in-Publication Data

Pain and emotion in modern history / edited by Robert Gregory Boddice
(assistant professor of history, Freie Universität Berlin, Germany).

pages cm

Summary: "Pain and Emotion in Modern History is a rich exploration of the affective
expression of pain, the emotional experience of pain, and the experience of others' pain
as pain. Drawing on the expertise of historical, literary and philosophical scholarship,
practising physicians, the medical humanities, and conceptual artists, this is a true
interdisciplinary collaboration, styled as a history. It explores pain at the intersection
of the living, suffering body, and the discursive cultural webs that entangle it in its
specific moment. This volume goes beyond the typical spaces and parameters of pain,
from the operating theatre to the waiting room; from the moment of birth to its
anticipation and aftermath; from the body in pain to the body in a culture of pain.
Most importantly, it moves from the narrowly physical to the broadly emotional,
enabling the enrichment of the medical history of pain, as well as setting a new
agenda for medical history"—Provided by publisher.

Includes bibliographical references and index.

1. Pain—History.   2. Pain—Social aspects—History.   3. Pain—Psychological
aspects—History.   4. Emotions—History.   5. Suffering—History.   6. History,
Modern.   7. Social history.   8. Medicine—History.   I. Boddice, Rob.
RB127.P3321225   2014
152.1'82409—dc23   2014019518

Typeset by MPS Limited, Chennai, India.

Transferred to Digital Printing in 2014

Contents

List of Figures

Acknowledgements

Joanna Bourke invited me to be a 'Fellow in Pain' at the Birkbeck Pain Project in London in 2012. Without that invitation, this book would not exist. My greatest debts are to Joanna, to the brilliant support from Carmen Mangion and Louise Hide at the Pain Project, to the Birkbeck Institute for the Humanities, and to the Wellcome Trust, who made my fellowship possible. The fellowship culminated in a public conference on the theme of 'pain as emotion, emotion as pain' and brought together many of the scholars, artists and physicians who now appear in this book. I thank them all for facilitating a painless conference and editing experience.

Further acknowledgements are due to the series editors, William Reddy and David Lemmings, and to Jenny McCall and Holly Tyler at Palgrave Macmillan, for encouraging this project from the start. I owe a debt too to the institutions that have supported the work I have been able to do in preparing this book: the Dahlem Research School, the Excellence Cluster Languages of Emotion and the Friedrich-Meinecke-Institut at Freie Universität Berlin, and the Centre for the History of Emotions at the Max Planck Institute for Human Development, Berlin, under the careful stewardship of Ute Frevert. The project has been generously funded by a Marie-Curie/Cofund award and a DFG grant. Thanks also to Thomas Dixon and the Centre for the History of Emotions at Queen Mary,University of London for generously entertaining some of my thoughts concerning 'the other body in pain'. Additional thanks go to the Wellcome Library, London and to Johanna Willenfelt for their generous supply of images.

I have incurred a number of personal and professional debts along the way and must make special mention of the following: Otniel Dror, Daniel Goldberg, Martin Lücke, Javier Moscoso and Jan Plamper. Last but by no means least, thanks to Stephanie Olsen, who always gets me through these things with the thought 'It didn't hurt a bit'.

Notes on Contributors

David Biro is Associate Clinical Professor of Dermatology at SUNY Downstate Medical Center in Brooklyn, New York. He is the author of *One Hundred Days: My Unexpected Journey from Doctor to Patient* (2000) and *The Language of Pain: Finding Words, Compassion, and Relief* (2010).

Rob Boddice holds a Deutsche Forschungsgemeinschaft grant for the project 'The Science of Sympathy: Morality, Evolution and Victorian Civilisation' at the Friedrich-Meinecke-Institut, Freie Universität Berlin. He is also Research Fellow at the Centre for the History of Emotions at the Max Planck Institute for Human Development, Berlin. He has published widely in the field of human-animal relations, the history of cruelty, the history of science and the history of masculinities.

Joanna Bourke is Professor of History in the School of History, Classics and Archaeology at Birkbeck College. She is the author or editor of ten books, including the Wolfson Prize-winning *An Intimate History of Killing: Face-to-Face Killing in Twentieth-Century Warfare* (1999) and most recently *What it Means to be Human: Reflections from 1791 to the Present* (2011). Her new book, *The Story of Pain*, will be published by Oxford University Press in 2014.

James Burnham Sedgwick is Assistant Professor of History at Acadia University in Nova Scotia, Canada. Currently being revised, his doctoral work captured the involute meaning and experience of 'being international' in the modern age by reinterpreting the trial of Japan's wartime leadership as both a groundbreaking judicial undertaking and a pioneering multilateral institution.

Sheena Culley is a PhD candidate at the London Graduate School, Kingston University. She is exploring the idea of comfort from 1850 to the present day, taking ideas from philosophy, psychoanalysis, material culture and neuroscience. Her research interests encompass aesthetics, emotion, body image, subjectivity and habit, and her academic contributions aim critically and theoretically to engage with overlooked aspects of everyday life.

Liz Gray is a PhD candidate at the Centre for the History of the Emotions at Queen Mary, University of London, examining the developing ideas of comparative psychology through the nineteenth century, in particular the work of William Lauder Lindsay (1829–80) and the use of animals within research on the science and pathologies of the mind.

Daniel J.R. Grey is Lecturer in World History at Plymouth University. He is currently working on two books: *Degrees of Guilt: Infanticide in England 1860–1960* (Liverpool University Press, forthcoming) and *Feminist Campaigns against Child Sexual Abuse: Britain and India 1860–1947* (Continuum Press, forthcoming).

Javier Moscoso is principal investigator of the HIST-EX research group on the history and philosophy of emotional experience and Research Professor of History and Philosophy of Science at the Spanish National Research Council. He was the curator of 'Pain. Passion. Compassion. Sensibility' at the Science Museum in London in 2004. He has recently published, both in English and Spanish, *Pain: A Cultural History* (Palgrave Macmillan, 2012).

Linda Raphael is Director of Medical Humanities in the School of Medicine at George Washington University. She is the author of *Narrative Skepticism: Representations of Consciousness in Fiction* (2001) and *When Night Fell: An Anthology of Holocaust Short Stories* (1999).

Danny Rees is Engagement Officer for the Wellcome Library, a global collection of material relating to the history of health and medicine, where he delivers public lectures on the history of physiognomy.

Paolo Santangelo is Professor of East Asian History at Sapienza University of Rome. He directs the book series 'Emotions and States of Mind in East Asia' (Brill) and as been editor of the annual publication *Ming Qing Yanjiu/Studies* since 1992. The author of many books in Italian, English and Chinese, his latest is *Laughing in Chinese. Emotions behind Smiles and Laughter: From Facial Expression to Literary Descriptions* (2012).

Noémi Tousignant is Research Associate in the Anthropologies of African Biosciences research group at the University of Cambridge. Since completing her PhD on pain-measuring practices in the USA, she has published articles in the *Journal for the History of Medicine and Allied Sciences* and *Social History of Medicine*. She is currently writing a book on the intersections of poisons and progress in postcolonial Senegalese science.

Johanna Willenfelt is a visual artist and PhD candidate in Fine Art at the University of Cumbria. Since receiving her MFA in 2010 (Valand School of Fine Arts, Gothenburg), she has exhibited individually in museums as well as in hospital environments. Her practice-based research negotiates embodied knowledge in pain relations, the interpersonal and beyond.

Wilfried Witte practises internal medicine and anaesthesiology, and is currently a specialist (*Facharzt*) in anaesthesiology at the Charité-University Clinic of Berlin. He is developing new research projects on the history of encephalitis lethargica and the history of pain management at the Institute for the History of Medicine in Berlin.

Whitney Wood is a PhD candidate at the Department of History at Wilfrid Laurier University in Waterloo, Canada. Her doctoral project, funded by the Social Sciences and Humanities Research Council of Canada, is a study of medical and cultural attitudes towards women's bodies, pain, and childbirth in late-nineteenth and early-twentieth-century English Canada.

1
Introduction: Hurt Feelings?

Rob Boddice

Figuring pain

Pain and Emotion in Modern History is a rich exploration of the affective expression of pain, the emotional experience of pain (with or without lesion) and the experience of others' pain as pain (sympathy, compassion, pity, tenderness).[1] Immediately it should strike the reader that 'pain' is at best a confusing label; at worst it is hopelessly inadequate. It must describe, at one and the same time, an appearance or surface, an inner (physiological and neurological) state and the reception of both these things as they are projected by another. This book goes a long way towards unpacking the polyvalence of 'pain'. Essential to this project is a conviction that pain, in all its complexity, has been, is and can be expressed – bodily, orally, emotionally and linguistically.

What Elaine Scarry described as the inexpressibility of pain in her seminal work *The Body in Pain* understated human capacities for articulating their suffering on the one hand and implicitly overstated human capacities for articulating all other emotions on the other.[2] Jerome Kagan has highlighted 'the inadequacy of most languages to capture the range of intensity and quality of frequent human experiences', whether an experience is of joyous guilt or of angry shame.[3] We translate emotions into verbiage imprecisely. Insofar as pain is an emotion – something for which the authors in this book argue strongly – then we translate that experience imprecisely too. But imprecision, clearly marked by an acute awareness that it is imprecise, tends only to heighten human efforts to enrich or deepen the confusion. As we translate bodily experience into words, grimaces and art – as we make metaphors of our inner experiences – we literally 'figure out' what we feel.[4] These figures may lack definition, but they are no less evocative

1

for that. And just as I 'figure out' how I feel, so my witness reads my figures, checks them against her own and, to some degree, understands. The degree of success in that process will depend on her knowing me, my context, my history and on our sharing a culture. As Kagan reminds us, 'reactions of living things are always affected by their past'.[5] The meaning is perhaps even more profound than he intended. For it is not just our individual or immediate pasts that affect us; the pasts of our families and societies do so too. We are, insofar as we are translators of biological states, cultural historical biological beings. It should go without saying, therefore, that what we *make* of pain – how we translate states of suffering – is also dependent on time and place.

Kagan can be usefully employed to clarify this point. 'It is critical', he says, 'to distinguish between a feeling, which is a perception, and a semantic concept.'[6] Feelings may be held in common, but they are described in different ways. In naming, in translating, we endow meaning. Even if emotions are, physiologically, changeable only over evolutionary time, what they mean changes far more rapidly than that. The biological sciences are not usually concerned with what things mean. Much of the squabbling over what constitutes pain and what constitutes emotion can be reduced to a perception among humanities scholars that scientists are confusing what emotions and pain *are* with what they mean, while scientists similarly suspect the humanities of confusing what emotions and pain *mean* with what they are. The nub of this problem is that we all share the labels 'emotions' and 'pain', but our referents are of entirely different orders. Kagan's call for 'a moratorium on the use of single words' for emotions, and instead to 'write about emotional processes with full sentences rather than ambiguous, naked concepts', is at the heart of what this book does.[7] While the authors all refer back to common labels – 'pain', 'emotion', 'suffering', 'sympathy' – they do so within a disciplinarily attuned context, leaving little room for confusion. We should have no problem admitting, between the humanities and the natural sciences, that we are often not talking about the same things, but merely about different components of an holistic knowledge, the parts of which must be commensurate if we are all to stand behind the respective truths of each discipline. We are blighted by a common language, but the solution lies in the training of the ear and of the pen.

Contexts of emotion

The investigation here concerns the emotional context of different kinds of pain and the development of an emerging concept of physical pain as intrinsically emotional/affective. The book goes beyond the typical spaces and parameters of pain, from the operating theatre or clinical office to the waiting room; from the moment of birth to its anticipation and its aftermath; from first-hand testimony to its reception; from the body in pain to the body in a culture of pain. Most importantly, it moves from the narrowly physical to the broadly emotional. The emotional component enables the enrichment of the medical history of pain, as well as setting a new agenda for medical history. Grief, anxiety, depression, hysteria, nervousness, despair and other 'mental illnesses' fall within its scope, and the historical mutability and efficacy of placebo or the 'moral economy of hope' (Moscoso) is given its due in the story of fighting pain. *Pain and Emotion* attempts, as Joanna Bourke puts it in this volume, 'to distinguish the experience of pain from the pain of experience'. From judicial courts hearing war crimes evidence to the parent in the paediatric hospital, the book also deals with the problem of pain caused by bearing witness to, or by the imagination of, the pain of others.

To think of the Other in pain is a useful way of getting to the heart of human cultures of pain; to understand the value of pain, the fear of pain, and the stimuli to pity, tenderness, compassion and sympathy, all of which historically and literally have denoted the emotional pain, some of it enjoyable, of the *witness* to pain. Humans are involved in a never-ending process of bearing witness to pain or of choosing not to bear witness.[8] The pain of infants, the pain of women, the pain of racial others, dubbed 'inferiors', the pain of the poor or uneducated, dubbed 'degenerate', the pain of the mentally ill, the pain of the old: each in its own way has been othered, sidelined, reduced, justified, condoned, condemned and mythologised. A long-running notion asserted that the more highly organised a being, the more sensitive to pain it would be, which has led on the one hand to a number of ingenious metrics for determining how organised an organism might be and on the other hand to an archive of apologies and excuses for pain, on the basis that some sufferers did not really suffer at all.[9]

The acknowledgment that pain is at least in part emotional is not new, but exactly what this means is still shrouded in uncertainty, and the implications for the medical definition and clinical approaches to the treatment of pain are in the relatively early stages of being worked out. The International Association for the Study of Pain's definition of pain already contains this emotional element and a significant nod towards metaphor, where pain is 'an unpleasant sensory and emotional experience associated with actual or potential tissue damage, or described in terms of such damage'.[10] We can say with certainty that pain can no longer be reduced only to the location and management of a lesion, and we can admit the neurological existence of pain in the brain; we can, in a rudimentary way, authenticate the experience of pain without bodily injury and begin to approach an understanding of the emotional component of what makes bodily injury hurt.[11] As David B. Morris has pointed out:

> postmodern pain is no longer officially divided along Cartesian lines and put into separate boxes labeled physical and mental. It is ... a biocultural event, subject to both individual and transpersonal modulation, as variable across social boundaries as other events with mixed biological and cultural significance ... Pain ... has been remade as irreversibly open to the hubbub of human social and psychic life and thus open to an inescapable intersection with narrative.[12]

But how far has that openness penetrated and how far have its implications been apprehended? Lisa Folkmarson Käll introduces a recent multi-disciplinary intervention in pain studies with the following assertion: 'Pain is the source of sorrow, suffering, hopelessness and frustration.'[13] This book, which in some ways complements that of Folkmarson Käll, takes more than merely semantic issue with this assertion, arguing forcefully that pain *is* sorrow, suffering, hopelessness and frustration. Put another way, physical pain is not meaningful without some or all of these things or without some other affective component (even pleasure, joy or ecstasy). Nor is pain, insofar as the experience is concerned, really conceivable without these affective components. This book explores the historical failure to recognise the pain that is invisible to the observer, however plain it is to the sufferer; the pain that has been reduced to mere 'suffering', an emotional quality made distinct from pain. To understand, both scientifically and culturally, that suffering *is* pain and that pain *is* emotion – that feelings *hurt* – is vital if we are to relocate the human experience of pain and if we are to understand the violence of denial, of pain that doesn't *count.*

Politics and meaning

The unifying factor among the authors here is the politics of pain, for always in question is the reason why one person's pain is important and treated, while another's is unimportant and ignored. How and why have sufferers been excluded from the realm of sympathy (and medical science) on the basis that their pain was merely an emotional problem (the shell shocked or the 'hysterical', for example) or because of a presumed dullness of nervous excitation or emotional capacities (babies, animals, 'savages'), or, conversely, included precisely because of heightened emotional sensitivities (the civilised, the feminine)? To begin to answer these questions entails putting pain into cultural context, where the experience of the person in pain relates to socially instituted and culturally embedded systems of inclusion and exclusion, along lines of gender, age, class, race and species.[14] Pain is only pain where it accords with tacit rules for the acceptable expression or the experience of pain. And to this end, the question of whose pain is *authentic* is a question of power. It is also a question of history, since constructions of authenticity never attain universal status. In historicising and deconstructing pain, and by revisiting those historical actors whose pain was valid and those who were *invalid, Pain and Emotion* not only sets a stark agenda for a new history of pain, it also sends a message to contemporary pain specialists, pain clinicians and pain managers to take a reflexive approach as to why pain is considered to be *this* but not *that, here* but not *there*, in the present. The purpose is not to redefine pain, but to lay bare its politics, ever changing and more or less subtle, but ever-present and always to someone's disadvantage.

In order to understand pain and to give pain a history, we are automatically pursuing a cultural approach to pain; we are asking 'what does, or what did, pain *mean?*', both for sufferers and for the communities in which suffering takes (or took) place. To ask this question has the effect of appropriating the definition of pain from physiological and neurological descriptions of what pain is, which are really descriptions, or depictions, of how pain works in the body and in the brain. The shared fundamental premise of the contributors to this book is that without an understanding of what pain means, any amount of information about how pain works will not get us very far in our ability to understand, let alone manage, pain. Broadly, we follow the observation of Richard McNally:

A functionalist definition of pain would specify its input (for example, tissue damage), its output (for example, grimacing, verbal complaints, reaching for a bottle of aspirin), and its relation to other states. We might elucidate the neurophysiologic processes occurring as a person is writhing in pain, including performing functional magnetic resonance imaging (fMRI) of the person's brain, and we might even program a robot to mimic pain. All of these procedures yield objective, publically observable data accessible to any third party. But none disclose [*sic*] what it is like for the person who is experiencing the pain itself.[15]

Therefore, we collectively ask: 'What does pain signify?' To look for meaning, for significance, leads us to the context of emotions. We find the meaning of pain in its expression, and whether that expression is a scream, a supressed scream, a stoic refusal to surrender to pain or a literary inquiry into the ravages of it, we find that pain is manifested in affective ways. Our encounter with pain, especially with the pain of others, is an encounter always with emotions, and it takes place always within a culture of pain: a limited, prescribed and unconsciously followed set of conventions that set out how, why and when pain can be expressed.

William Reddy's seminal work on the history of emotions has done the most to develop a relationship between new research in human neurobiology and theories of social construction of emotions, which had previously inhabited seemingly intractably incommensurate points of view.[16] The idea that our emotional utterances are acts of translation, processes of negotiation of how we feel (our biology) with how we are expected to express that feeling (our culture), surely has huge implications for the study of pain, especially when the emotions are now recognised as such an important component in the experience of pain. Indeed, Reddy's claim that these processes – what he calls 'emotives' – are always to some extent a failure might help explain, in David B. Morris' terms, just why pain can be such a 'mystery'.[17] Reddy demonstrates that the cultural delimitation of acceptable forms of emotional expression never quite match how we feel. We attempt to match our feeling to what is expected. The closer we get to parity between feeling and feeling rule, the better. But there is never exact parity, and more extreme failures are common. These emotive failures, Reddy shows, can lead to personal and even societal crises, where outpourings of 'inappropriate' emotional expressions seem to challenge the 'natural' order of things. The history, for example, of phantom limb pain, detailed in different contexts in this book by Joanna Bourke and Wilfried Witte, surely fits this description. A patient's pain, not fitting accepted medical models of how and why pain comes about, compounds the patient's despair. Unable to convince the medical establishment of the legitimacy of the complaint, expressions of pain seem to transgress what is acceptable or explainable. The pain does not fit.

As Javier Moscoso has pointed out, 'there is no suffering that does not entail a social appraisal and, by extension, a form of expression linked to cultural guidelines and expectations'.[18] These are 'pain rules' or 'pain styles'.[19] Just as with the expression of other emotions, *context*, by limiting the affective range of how we *ought* to feel, has the effect of influencing how we *do* feel. *Context defines experience.* And just as with other emotions, the personal experience of pain often fails to fit neatly into these cultural frameworks, leading perhaps to the transgression of social or moral rules, to increased anguish, to reactionary measures from the society that does not, will not or cannot understand the pain to which it bears witness and, sometimes, to conciliatory measures that open up society to new understandings of pain. The history of emotion as pain opens up a history of the denial of pain, as well as grounding a narrative of change that can make sense of how and why certain kinds of pain are no longer denied. In these ways, *experience also defines context.*

Bioculture

The importance of emotions in the experience of pain is magnified once it is granted that the emotions themselves are culturally and historically specific, rather than universal. A parallel can be drawn using Melzack and Wall's description of the cultural contingency of pain perception. Having established a universal register of pain sensation, Melzack and Wall nevertheless show that the perception, experience and tolerance of pain are culturally contingent. What a 'painful' experience means depends on the meanings assigned to it by a given culture and by the past experiences of the individual currently undergoing a 'painful' experience. The body, in other words, is a cultural body. An individual's physiology is in a dynamic relationship with where it is and what it means to be there. Recent research on the emotions has more or less reached the same conclusions.[20] Regardless of whether or not universals might be found in the visceral response of the body to stimuli, emotions happen contextually.[21] Expression is a dynamic process of negotiation between inward feelings and outward expectations, norms, taboos and so on. If we grant that pain experience is affective and that emotions are historical, then differences in pain experience can in part be explained by the history of emotions. We cannot simply look to fMRI images that detail the activation of affective centres of the brain in pain experience. These alone only tell us that affect is indeed integral to pain experience. But since we can only know the meaning of affective experience in historical and cultural context, it is necessarily to history and to culture that we must look to understand it.

A richer account of the history of pain must emerge from this focus, but there is also potential for clinical approaches to pain to shift ground. While contemporary definitions of pain have acknowledged its emotional component, with consequences for palliative care, the degree of sensitivity to the mutability of 'emotions' has been low. This cannot be written off simply as a postmodern critique of universalist tendencies within psychology and neuroscience.[22] It is an observation of bodies in context, of experience through culture. As Melzack and Wall put it, 'pain is not simply a function of the amount of bodily damage … Even the culture in which we have been brought up plays an essential role in how we feel and respond to pain'.[23] If how we feel is in part explainable only by the 'feeling rules' of the culture we inhabit, then attempts to fix diagnostic labels onto emotional states – 'hysteria', 'depression', 'neurosis', 'anxiety' – in order the better to treat patients will automatically lack the reflexive cultural sensitivity required to understand why a person may feel the way she feels in a given painful experience. If psychological approaches to pain employ fixed emotional categories, then their efficacy might only go so far and only last so long.

Nikola Grahek convincingly establishes a distinction between 'feeling pain' and 'being in pain' in a bid to demonstrate the richly entangled phenomena of a painful experience beyond the mere physical registration of injury. He quotes Wall to good effect: 'I have never felt a pure pain … A particular pain is at the same time painful, miserable, disturbing, and so on. I have never heard a patient speak of a pain isolated from its companion affect.'[24] In order to make his point, Grahek documents the rare cases of patients whose affective responses to pain have been cut off, either surgically or congenitally (pain asymbolia). These people feel pain, but experience none of its unpleasantness (and are therefore in constant, unappreciated danger). Pain ceases to mean anything. Without *suffering*, a term that awkwardly captures the affective element of pain, physical pain is reduced to an objective state. It serves no purpose, removing the fear from danger. Imagine the classic Cartesian image of the figure with his foot in the fire, demonstrating the mechanics of reflex action. The bell rope in the foot is pulled and the bell in the brain does indeed ring, but it is after all only a bell. The foot remains in the fire, to the complete indifference of the man. The man machine simply does not work without its corresponding affective package.[25] To sum up the complexity of the experience of being in pain, Grahek employs an essential language of emotions:

> pain, when stripped of [affective, cognitive, and behavioural] components, loses all its representational and motivational force. It comes to nothing in the sense that it is no longer a signal of threat

or damage for the subject, and doesn't move (*emotio*) his mind and body in any way. If that is the case, it follows that the basic representational and motivational force of pain should not be sought in its sensory components, but rather in its affective, cognitive, and behavioral dimensions.[26]

While this may be increasingly recognised within medical understandings of pain, the contribution of this book lies in its destabilisation of what we mean by the 'affective dimension'. If we acknowledge that the experience of pain is bound up with, amongst other things, our specific fear of what pain may signify, then the historical context of fear will alter pain experiences. A pain in the chest and left arm may make us fear we are having a heart attack, but what if we do not know what a heart attack is? What if we think our pain is visited upon us by a vengeful God, in consequence of sin, or by witchcraft? Is the pain experience of a heart attack, as 'we' now commonly understand it, likely to be the same? The consequence of historicising the emotions is that palliative care directed at reassurance – that a patient may gain control over his or her pain – must have to do with contingent contexts of fear, anxiety, depression and so on. The sentence beginning 'It will be alright, because ...' is always completed by a culturally relative statement of reassurance (even where that reassurance takes the form of medical knowledge of how to fix the problem).

David B. Morris is perhaps the chief advocate of the importance of emotions in the experience of pain, especially but not exclusively in chronic pain. In *The Culture of Pain*, Morris imagines a carpenter 'who slips from a high roof and lands on his toolbox'. This doubly unlucky fellow 'will suffer major injuries', but the physical damage done by the fall (or rather by the landing) will be compounded by a complex pain that 'may also enfold deep anxieties about whether he will ever work again, support his family, and lead a normal life'.[27] I acknowledge the truth of this, but also recognise the need to locate, culturally, historically and politically, those anxieties, the patriarchal attitude and the concept of normality.

Morris has forcefully and persuasively argued that: 'Pain is always personal and always cultural. It is thus always open to the variable influx of meaning.' Implicit to this claim is that pain is also historical, and indeed the historical method still provides the best way of describing and explaining changes to personal and cultural contexts of meaning. As Morris himself states: 'Meaning ... is something that exists only within the shifting processes of human culture and of individual minds.

It therefore can never successfully shut itself off from change.'[28] This insight can be taken further once we acknowledge that our interpretive encounter with pain begins at the level of experience.

Interdisciplinary dialogue

In order fully to represent the contextual contingency of emotional pain experience, this book draws on the expertise of historical, literary and philosophical scholarship, practising physicians, the medical humanities and conceptual artists. Styled as a history, the book is nevertheless a true interdisciplinary collaboration. Its geographical scope includes North America, Europe, Britain, Japan and China with a temporal range from the late seventeenth century (or earlier in the case of China) to the present. Unlike many multi-disciplinary projects, the attempt here was to have different disciplines deeply engage each other. This began as a listening exercise and as a talking exercise at the Birkbeck Pain Project in London. The critical dynamic was to have medical practitioners engage with humanities scholars and vice versa, for the subject of pain lies at the intersection of the living, suffering body, its neurology and its 'wiring', and the discursive cultural webs that entangle it in its specific moment. To understand pain (and presumably, therefore, to treat it) requires biological knowledge and cultural reflexivity.

The book that has developed from this discussion preserves the engagement across disciplinary lines, in some ways disrupting expertise about pain and in other ways meaningfully enhancing our understanding. The volume begins with three narratives that demonstrate the distance between understandings of pain over time and place. Javier Moscoso's story of curious cancer remedies, hope and placebo is at once revealing of the strangeness of past medicine and of its continuing relevance for present treatment. Paolo Santangelo demonstrates an intertwining of physical and emotional experience to reveal how pain was perceived in late Imperial China, emphasising the importance of metaphor and affective practices that substantiate pain experiences. David Biro also recognises the importance of metaphor, but in a context far removed from late Imperial China or early modern Europe. Couching his study in the contemporary history of medical and neuroscientific understandings of pain from the perspective of the medical practitioner, he demonstrates that emotional 'pain' caused by, amongst other things, social exclusion, rejection or grief is of a piece in the brain with pain caused by physical injury. The sequential placement of these chapters disrupts any easy understanding of pain. On the one hand, they are far apart from each other, demonstrating the massive range of possibilities for the communication and alleviation of pain; on the other hand, they are strikingly close to one another, demonstrating a continuity in the devices employed to understand, translate and treat pain. These chapters, along with this introduction, set the agenda for the specific case studies that follow.

In particular, the subsequent chapters attempt to situate the experience of pain within the contexts of distinct social and cultural practices that prescribe the limits of meaningful suffering. This brings into play the institutions of medical knowledge, medical practice and the pharmaceutical industry, situating them according to their capacity to define, affirm or deny pain. Chapters 5 and 6 are natural companions. Joanna Bourke's essay on the suffering associated with phantom limb is a microhistorical account of one man's experience of a lifetime in pain, struggling to gain acknowledgment and relief from a system that was structurally unable or unwilling meaningfully to offer either. It enters into a dialogue with Wilfried Witte's account of the same subject from the point of view of the kinds of medical institutions and practices that struggled to hear and to understand the experience of a pain whose cause was literally invisible. This juxtaposition of cultural historian (Bourke) and anaesthesiologist (Witte) casts a raking light over the history of medicine that helps us better to see institutional power dynamics and their effects on the sufferer. For the cultural historian, Witte's insight is politically significant; for the anaesthesiologist, Bourke's narrative is clinically significant.

Continuing with these themes, Noémi Tousignant focuses on the sponsorship of pain-measuring practices during the twentieth century in the USA. She argues that the intensification of efforts precisely to correlate 'suffering' with physiological indicators, along with the development of promising analgesic substances, gave rise to increasingly emotion-based definitions of pain. By way of contrast, Sheena Culley explores the extent to which common analgesics were marketed after the fashion of nineteenth-century nostrums, appealing especially to fraught housewives whose nerves were thought to be shattered by housework. The highly gendered quackery of the pharmaceutical industry is given an ironic twist by recent research that suggests, in accord with Biro's evidence of the role of the brain's affective centres in pain experience, that common painkillers might indeed be effective in relieving 'social pain'.

Liz Gray and Danny Rees both explore the complex interpretations of expressions of pain, predominantly in the context of nineteenth-century science. Through a biographical account of the physician William Lauder Lindsay, Gray demonstrates the differences of opinion on the meaning of the 'writhing' of animals. The differences centre on how pain works and on the resolution of one of modernity's classic problems, that of mind-body dualism. Rees shifts the focus to humans and to the pictorial history of representations of the human face in pain or, rather, not in pain. Most of Rees' images depict manufactured expressions, made without pain, but in the name of proving a universality of human emotional expression. The images in fact make a case for the difficulty of ever authenticating an emotion by its physical sign.

Whitney Wood and Daniel Grey deal with the emotional pains of motherhood, antenatal and postnatal, respectively. Wood argues that the perception of civilised women, even among women themselves, as too 'sensitive' and too overwhelmed by fear to bear the pain of childbirth led to the increasing pathologisation and medicalisation of pregnancy and childbirth. Grey, examining a number of cases of maternal infanticide, discusses the perceived relationship of physical pain and mental anguish as a cultural explanation for the crime. He documents the cultural response to infanticide as a history of compassion, of a medical, legal and judicial witnessing of the 'agony of despair', steered by a growing conviction that 'certain' women could be led to a state of temporary dissociation by a combination of physical and emotional pain.

This perspective sets up the book's conclusion, which encompasses three chapters that explore various disciplinary approaches towards the experience of another's pain, which the chapters by Gray and Rees also touch upon. Linda Raphael turns to fiction. The two stories on which she focuses both involve struggles to understand, interpret and act upon the pain of others, but it is the reader of such stories that Raphael has in mind. Not only does fiction draw us into the complexities of sympathy and vicarious pain, it offers us experience and insight into the 'heterogeneous emotional responses' aroused by 'regarding pain'. Such responses are examined by James Burnham Sedgwick in the context of International Military Tribunals. Using a notion of vicarious trauma, he documents the 'traumatic growth' of two individuals who bore witness to the pain of others at the trials in Nuremberg and Tokyo. Finally, Johanna Willenfelt employs research-based visual art to arrive at a new understanding of sharing pain. Drawing heavily on 'object-oriented ontology', a philosophy that endows objects with agency, and the psychological inspiration of William James, Willenfelt attempts to grasp pain as an 'emotive, almost tangible object, in a world replete with other objects'.

Taken together, the chapters in this book demonstrate that the more we zero in on pain, the more its slipperiness is revealed. If we admit that pain is historical, cultural and emotional, then there can be no final word on pain, either what it is or what it means. But the book also demonstrates that pain is all the more approachable, translatable and treatable once its contingencies are understood. This in turn demands that those who have to deal with pain, whether through scholarship of one kind or another or in a clinical setting, always deal with it reflexively, critically and contextually. The mystery of what pain is for individual or collective sufferers can be unlocked by coming to terms with the experience of suffering, of bodies in culture, of feelings in context.

Notes

1. On affective expression of pain, Javier Moscoso has set the bar: *Pain: A Cultural History* (Basingstoke: Palgrave Macmillan, 2012). See also Esther Cohen, Leona Toker, Manuela Consonni and Otniel Dror (eds), *Knowledge and Pain* (Amsterdam: Rodopi, 2012). On the historical problem of pain without lesion, see Andrew Hodgkiss, *From Lesion to Metaphor: Chronic Pain in British, French and German Medical Writings, 1800–1914* (Amsterdam: Rodopi, 2000); and Daniel Goldberg, 'Pain without Lesion: Debate among American Neurologists, 1850–1900', *19: Interdisciplinary Studies in the Long Nineteenth Century*, 15 (2012). Few works on sympathy have explored it as vicarious suffering. Max Scheler's classic *The Nature of Sympathy* (trans. Peter Heath) (London: Routledge & Kegan Paul, 1954) still holds up. David Konstan's *Pity Transformed* (London: Duckworth, 2001) is exemplary in its defamiliarisation of pain and compassion. See also Moscoso, *Pain*, Chapter 3; and Esther Cohen and Leona Toker's introduction to *Knowledge and Pain*.
2. Elaine Scarry, *The Body in Pain* (Oxford University Press, 1987).
3. Jerome Kagan, *What is Emotion? History, Measures, and Meanings* (New Haven: Yale University Press, 2007), 9.
4. See David Biro, *The Language of Pain: Finding Words, Compassion, and Relief* (New York: Norton, 2000).
5. Kagan, *What is Emotion?*, 8.
6. *Ibid.*, 24.
7. *Ibid.*, 216.
8. See Susan Sontag, *Regarding the Pain of Others* (New York: Picador, 2003).
9. The capacity to suffer meaningfully has been bound up with what it means to be human and who gets to be included in that category. See Joanna Bourke, *What it Means to be Human: Reflections from 1791 to the Present* (London: Virago, 2011), 65–123; see also Lucy Bending, *The Representation of Bodily Pain in Nineteenth-Century English Culture* (Oxford University Press, 2000), 177–239.
10. Adapted from 'Part III: Pain Terms, A Current List with Definitions and Notes on Usage', in H. Merskey and N. Bogduk (eds), *Classification of Chronic Pain*,

2nd edn (Seattle: IASP Press, 1994), 209–14. See www.iasp-pain.org, under resources, taxonomy (date accessed 21 February 2014).

11.   See David Biro, Chapter 4, this volume. See also Jaak Panksepp, *Affective Neuroscience: The Foundations of Human and Animal Emotions* (New York: Oxford University Press, 1998), 187–224, 261–79; Giacomo Rizzolatti and Corrado Sinigaglia, *Mirrors in the Brain: How Our Minds Share Actions and Emotions* (trans. Frances Anderson) (Oxford University Press, 2006); Bruce M. Hood, *The Self Illusion: How the Social Brain Creates Identity* (Oxford University Press, 2012), 187–91.

12.   David B. Morris, 'Narrative, Ethics, and Pain: Thinking with Stories', *Narrative*, 9 (2001): 55–77 at 60.

13.   Lisa Folkmarson Käll, *Dimensions of Pain: Humanities and Social Science Perspectives* (London: Routledge, 2013), 1.

14.   For an analogy, see Joanna Bourke, 'Fear and Anxiety: Writing about Emotion in Modern History', *History Workshop Journal*, 55 (2003): 111–33 at 124.

15.   Richard J. McNally, 'Fear, Anxiety, and their Disorders', in Jan Plamper and Benjamin Lazier (eds), *Fear Across the Disciplines* (University of Pittsburgh Press, 2012), 17.

16.   The most succinct expression of this is in William Reddy, 'Against Constructionism: The Historical Ethnography of Emotions', *Current Anthropology*, 38 (1997): 327–51. The full development of the idea is in William Reddy, *The Navigation of Feeling: A Framework for the History of Emotions* (Cambridge University Press, 2001). See also Rob Boddice, 'The Affective Turn: Historicising the Emotions', in Cristian Tileagă and Jovan Byford (eds), *Psychology and History: Interdisciplinary Explorations* (Cambridge University Press, 2014).

17.   David B. Morris, *The Culture of Pain* (Berkeley: University of California Press, 1991), 25.

18.   Moscoso, *Pain*, 56.

19.   There are various models of the social construction of emotions that map neatly onto pain. See Arlie Russell Hochschild, 'Emotion Work, Feeling Rules, and Social Structure', *American Journal of Sociology*, 85 (1979): 551–75; Barbara H. Rosenwein, 'Worrying about Emotions in History', *American Historical Review*, 107 (2002): 821–45; Peter N. Stearns and Carol Z. Stearns, 'Emotionology: Clarifying the History of Emotions and Emotional Standards', *American Historical Review*, 90 (1985): 813–36.

20.   Ronald Melzack and Patrick D. Wall, *The Challenge of Pain* (London: Penguin, 1996), 15–18. See also Jan Plamper, 'The History of Emotions: An Interview with William Reddy, Barbara Rosenwein, and Peter Stearns', *History and Theory*, 49 (2010): 237–65. For works that have put emotion in context, see Barbara Rosenwein (ed.), *Anger's Past: The Social Uses of an Emotion in the Middle Ages* (Ithaca: Cornell University Press, 1998); Barbara Rosenwein, *Emotional Communities in the Early Middle Ages* (Ithaca: Cornell University Press, 2006); Joanna Bourke, *Fear: A Cultural History* (London: Virago, 2005); Thomas Dixon, *From Passions to Emotions: The Creation of a Secular Psychological Category* (Cambridge University Press, 2003).

21.   On the history of the scientific quest to locate emotions in the body, see Otniel Dror, 'The Scientific Image of Emotion: Experience and Technologies of Inscription', *Configurations*, 7(3) (1999): 355–401.

22. See William R. Uttal, *The New Phrenology: The Limits of Localizing Cognitive Processes in the Brain* (Cambridge, MA: MIT Press, 2003); Paul W. Turke, 'Which Humans Behave Adaptively, and Why Does it Matter?', *Ethology and Sociobiology*, 11 (1990): 305–39 at 315, 321; Robert Turner and Charles Whitehead, 'How Collective Representations Can Change the Structure of the Brain', *Journal of Consciousness Studies*, 15 (2008): 43–57. The universalist position – that emotions (including pain) and their expressions are a constant among humans – is best exemplified by Paul Ekman. See Paul Ekman and Wallace Friesen, 'Constants across Cultures in the Face and Emotion', *Journal of Personality and Social Psychology*, 17 (1971): 124–9; Paul Ekman and Wallace Friesen, *Pictures of Facial Affect* (Palo Alto: Consulting Psychologists Press, 1976); Paul Ekman, *Emotions Revealed: Recognizing Faces and Feelings to Improve Communication and Emotional Life* (New York: Times Books, 2003). For a critique, see Boddice, 'Affective Turn'.

23. Melzack and Wall, *Challenge of Pain*, 15–18.

24. Nikola Grahek, *Feeling Pain and Being in Pain*, 2nd edn (Cambridge, MA: MIT Press, 2007), 36.

25. *Ibid.*, 31–6.

26. *Ibid.*, 40.

27. Morris, *Culture of Pain*, 158.

28. *Ibid.*, 25–6.

2
Exquisite and Lingering Pains: Facing Cancer in Early Modern Europe

Javier Moscoso

Introduction[1]

In 1781, José Flores, a medical doctor and Professor of Medicine at the University in Guatemala, published a small booklet on a new specific remedy for the cure of cancer.[2] This disease, already known to the ancient Egyptians, and described and named by Hippocrates, was very well known in the eighteenth century, despite its relatively low incidence.[3] Although according to Gaspard-Laurent Bayle, a medical doctor of the Charité Hospital, one out of seven patients who entered the Paris hospitals at the end of the eighteenth century died as a consequence of a form of this disease, other sources refer to a death rate of no more than 2 per cent at the end of the eighteenth century and around 2.4 per cent in 1840.[4] At the beginning of the twentieth century, the incidence of cancer reached a figure between 3 and 5 per cent, far away from tuberculosis, the cause of death of about 36 per cent of population.[5] At the time of the publication of José Flores' booklet, many doctors shared the impression that cancer, a deadly though relatively rare disease, was becoming increasingly frequent.[6]

According to Flores, José Ferrer, a Catalan who suffered from an incurable and terrible cancer, was inclined to try a local remedy employed by the Indians of the small village of San Juan de Amatitlán in the region of Sacatequez. It was there that the young doctor first came to know that some ulcers and cancerous tumours could be allegedly cured by eating the raw and palpitating flesh of local green lizards. With different variations and with the addition of new testimonies, Flores' text was republished again in Mexico in 1782 and in Spain in 1783. It was also later translated into French, English, Italian and German. Meanwhile, in 1800, the *London Medical Review* published a letter regarding the curative nature of the raw flesh of green lizards.[7]

16

This story of cancer and lizards may be read from different angles. First of all, it concerns the aspirations of a young physician searching for promotion within the medical profession of colonial Spanish America. It also relates to the tensions between local expertise and official knowledge. Finally – and this is what concerns us here most – it also refers to the degree of desperation involved in the experience of cancer patients. As in the case of other specific remedies that proliferated during the long eighteenth century, Flores' report refers to the stories of many other patients who recovered their health after chewing two, three or more lizards. Even if some of these patients were initially reluctant to masticate the raw flesh of these reptiles, they finally gave their consent, for, according to Flores, not so much effort was required to convince a hopeless patient to undertake some procedures.[8]

What the story of the lizards comes to suggest is that the efficiency or inefficiency of a given remedy may also depend on the way in which 'hope, after all, helps'.[9] As with many other specific remedies, perhaps patients did suffer, and eventually recovered, from other diseases and not from what today we consider cancer, but this would only call into question our historical or present understanding of the relationship between diseases, as socially determined nosological entities and illnesses, as the set of symptoms experienced by patients. In the case of cancer, as in any other 'open' conditions ('open' in the sense that the full understanding of the disease is still pending), the relationship between the subjective experience of the illness and the objective dimension of the disease remains blurred. On the one hand, unspecific and invisible symptoms, and pain in particular, helped to determine the presence of the disease. On the other hand, however, the way in which some of the symptoms were reported and felt depended on the social and collective ways in which cancer was described in medical and paramedical treatises. In both cases, the experience of cancer was framed within an emotional framework related to uncertainty, hope and fear. This emotional dimension of cancer and, more specifically, of cancer pain remains an historiographical issue,[10] for even if we know that all remedies suggested for the specific treatment of cancer were, from our own twenty-first-century perspective, totally inefficient, we also have the testimonies of our historical patients for whom those lizards became a painkiller, 'a more powerful painkiller than opium'.[11]

This chapter explores the relationship between the cultural and the natural history of symptoms. More particularly, it deals with the interplay between the objective evaluation of morphological signs and the subjective account provided by patients. It is not about the interplay between mind and body, or about the way in which the power of the mind may or may not produce a real cure in a given condition. I do not wish to take part in any quarrel about the reality or reach of the so-called 'placebo effect';[12] I only mean to shed some light on the historiographical issues to which the testimony of both patients and doctors give rise. While historians usually do not take part in debates related to the existence or non-existence of placebo effects, we may still have something to say regarding the way in which our sources should be interpreted and the form in which some of those illness narratives must be read. Anachronism may very well be one of the capital sins of our endeavours, but we are still forced to choose between the blind acceptance of our testimonies or the firm rejection of our historical records.

Cancer

Before cancer reached the 25 per cent death rate that it has today, the fear caused by the disease did not come from its statistical incidence. Cancer puzzled doctors and scared patients for three different reasons. First of all, it belonged to the category of 'incurable diseases'. Though the diagnosis of cancer was regarded by patients as a form of death sentence, the recognition of the incurability of the disease did not prevent the proliferation of remedies. On the contrary, from the mid-sixteenth century to the early 1830s, the general rejection of humoral theory led to an increase of competing explanations regarding its nature, causes and treatments. From the point of view of both patients and medical practitioners, experience came to show that cancer was not always deadly. Conversely, there were very good reasons to talk about the gradual or progressive incurability of the disease.[13] This was partly due to its elusive nature and, one may also argue, to a not inconsiderable number of incorrect or undetermined diagnoses. For Gamet, for example, cancerous affections included scrofula, scurvy and haemorrhoids, as well as hysterical, melancholic and nervous passions.[14] In an attempt to surpass the humoral explanation of cancerous growths, Enlightenment surgeons and physicians suggested many possible causes that could trigger the malady. According to Boerhaave, for example, a simple contusion could produce the worst tumour.[15] For Jean Astruc, a well-known surgeon and obstetrician, the willingness of some women to have their breasts handled could eventually generate that illness.[16] Louis Joseph Marie Robert included among the causes of this deadly disease almost anything that could count as a product of false vanity or pride: wine and liquors, coffee and spiced food, red and salted meat, the use of jewellery or perfumes, the enjoyment of music and dances, spectacles and games, vigils and noisy pleasures, the licentious paintings of the boudoirs, dissolute novels, erotic songs, walks and conversations with persons of the other sex, sadness, hatred and jealousy, idleness and celibacy.[17] In 1838, Dr Canquoin considered that, during his 13 years of medical practice, a more common cause of cancer was deep and chronic sadness.[18] He shared the ideas expressed by many other doctors who considered 'sorrows, annoyances and fears, anger or other negative moral impressions' as the most common moral causes of the condition.[19]

In the second place, cancer was for many a dangerous and contagious disease. While many surgeons compared it to aqua fortis, or to any acid factor that was present both in tumours and in the circulating blood, other physicians regarded it as a kind of putrefying alkali, a stagnant lymph or as a contagious virus. For Bernard Perylhe, for example, there was little doubt regarding its contagious nature, either in the form of emissions or through direct contact with cancerous tumours. This member of the Royal Academy of Surgery in Paris referred to the story of a man who, after having sucked the cancerous breast of his wife with the hopes of relieving her from her pains, soon after developed a cancerous affection of the lower part of the jaw, of which he died.[20] A similar misfortune happened to a Mr Smith, whose imprudent curiosity led him to taste the fluid proceeding from a cancerous gland, which he had just extirpated. The foetid and acid taste of this tumour continued to be so irremovably fixed to his tongue that an almost perpetual vomiting was brought on, and this was followed by marasmus.[21] When the first Cancer Hospital was opened in Rheims in 1740, this was in part due to the fear and apprehension that cancer patients produced in some other inmates. The alarm was so intense that the Hospital had to be moved outside of the city boundaries and, in 1779, a new building, the Hospital St Louis, was opened exclusively for cancer patients.[22]

Finally, there was no specific remedy, either to cure the illness or to alleviate its symptoms. The list of preparations employed for the cure or treatment of cancer comprised a huge variety of procedures, including the application of raw meat over the external cancerous tumour, to 'feed' the cancer and mitigate its hunger for human tissues. For many doctors and surgeons, all hopes of a cure rested on the possibility of separating the ill materials from the healthy body by means of suppuration, caustics, actual cautery or the knife.[23] In his *Treaty of Cancerous Affections*, Gamet mentioned, along with the above-mentioned lizards, hemlock, arsenic, and all the carbonates and muriate of iron, that is, all the different tinctures and compounds produced by the combination of iron oxide and hydrochloric acid. To these corrosive substances, he added muriate of barite, a highly toxic mineral, frequent bloodlettings and plenty of fresh air. The use of caustic and corrosive substances was in part reasonable, since cancer began to be considered as the result of a local affection rather than as the effect of a humoral imbalance. Though the efforts to eradicate the tumours were not always successful, the practice followed the logical path opened up through the new understanding of the disease. At the same time, the pains of the treatment, which seemed to imitate the gnawing and corrosive pains of the disease itself, suggested a correspondence between salvation and sacrifice. The application of corrosive substances onto the tumours was so painful than Canquoin classified them according to a scale of pain, which went from the less painful nitric acid to the unbearable pain produced by the application of antimony chloride.[24] In other cases, like in the use of the *alcali volaltil fluor*, the remedy was the same for both cancer and burns.[25] In the first decades of the nineteenth century, both Samuel Young and Joseph-Claude-Anthelme Récamier attempted to treat breast cancer by means of compression of the breast,[26] and from the eighteenth century until at least the 1820s, it was also treated with electricity.[27]

There were also, of course, 'secret remedies' and nostrums. As late as 1857, Weldon Fell of the University of New York mentioned various treatments for combating cancer, among them arsenic, jimson weed, hydrocyanic acid, certain animal substances such as cod-liver oil, belladonna poultices, mercury baths, carbonic acid, gastric juices, silver nitrate, zinc chloride and, of course, surgery. Similar to the case of the Guatemala lizard, Fell also proposed the use of a plant called *puccoon*, which he had heard about from the American Indians and which was considered an infallible cure.[28] If the lizards, the *puccoon* or the live toads that were supposed to suck out the cancer poison could be regarded as inefficient, those remedies found in the most serious treatises were not much different.[29] For those who believed in the acid nature of the cancer virus, the best prescription was alkalescent or alkaline medication. Martinet, for example, attempted a cure with the volatile alkali fluor (hydrofluoric acid), a highly corrosive substance. For others, the answer was ammonia. Le Febure suggested a cure based in the daily intake of a dose between one and six full spoons of arsenic – a remedy that led very often to the grave.[30] Professor Roux, for example, witnessed the death, seized by the convulsions and the most lively sufferings, of a young women whose breast had been severed and who had been treated thereafter with arsenic paste.[31] In his great compendium of tumours, Jean Astruc, one of the main physicians of the mid-eighteenth century, wrote of palliative remedies including the juice of houseleeks and nightshades (*solanum*), 'Saturn salt' (lead acetate, a highly poisonous substance), the brew or decoction of river crayfish, sea crabs, green frogs and snails. These latter could also be crushed and ground in a lead mortar, adding Saturn salt or liquor.[32] Prescriptions for external use included plasters of different kinds, nightshade, tobacco plants, arsenic paste in different compositions, oil of baked frogs, 'fresh veal or even a pigeon or any other warm-blooded animal cut up alive'.[33]

From the point of view of physicians and surgeons, cancer was discouraging; for patients, it was even worse.[34] To the rather confusing scenario regarding its aetiology, cancer sufferers also had to cope with the professional clash that surrounded the practice of medicine. Cancer treatment was a matter of concern not only for physicians and surgeons, but also for apothecaries, herbalists and charlatans. Though many medical treatises were inclined to discriminate between empirical and medical remedies, the possibilities certainly went much further than this classification tends to suggest.[35] The experience of cancer always involved an emotional experience and, more in particular, a correlation between fear and hope. Despite the reputation of the disease and its alleged incurability, remedies were always prescribed within the logical framework of patients' desires and healers' expectations.

Cancer pain

Regarding cancer, everything was doubtful, except pain. The prescriptions, the remedies, the expectations of patients, families and doctors were always insecure and uncertain, but pain was always there. Though rare, the disease was regarded with fear and apprehension, not only as a consequence of its apparent incurability (many patients did in fact survive), but because lancinating, darting and burning pains were among its most significant manifestations. 'Pain is perhaps the symptom which has, from the earliest recognition of cancerous diseases, attracted the greatest attention on the part of both patients and observers', wrote Walter Hayle Walshe in 1846.[36] Since not all tumours were considered cancerous, physical suffering became the key diagnostic element: 'one should consider pain as the specific and individual characteristic of cancer', wrote Jean-Baptiste Aillot in 1698.[37] It was pain, combined with the hardness of the tumour, that truly distinguished cancer, either internal or external, visible or hidden, from other indolent formations.[38] It was also pain that served to differentiate between benign and malignant cancer and to determine its incurability.[39]

From the point of view of its natural history, the development of the disease followed the same path as the unfolding of its symptoms. At first, there was only an insensitive growth, either an oedema or a scirrhus. Only when the scirrhus became painful were surgeons and physicians willing to accept the presence of a cancerous tumour.[40] For many of them, pain increased as the tumour grew.[41] The idea of cancer is terrible, wrote Houppeville, a surgeon from Rouen. The words *noli me tangere, loup, saratan, carcinoma* and *cancer* caused horror, not only because of the similarity of the cancerous tumour to a sea crab (a *cancre marin*), but because of the nature of cancer pain, which the French language referred to as *rongeante*, that is, the pain of corrosion and gnawing.[42] It was not the form, the swollen veins, that defined the disease, but the physical harm that it produced. It might look like a crab, like a crayfish, but what really mattered was that cancerous tumours got stuck onto the body and preyed on the flesh like crabs do, retaining whatever they had seized with their pincers.

The connection between pain and cancer was so tight that some physicians claimed to have distinguished cancerous tumours from other painful formations mainly on the basis of the nature and quality of the pain involved – for whereas cancer pain was gnawing, the pain of other soft tumours was pulsating. For Vacher, for example, the physical suffering associated with cancerous scirrhus resembled the pain caused by pins or needles.[43] For Bernard Peryrilhe, cancer pain was burning, pungent, lancinating or acrid.[44] Very often, patients described their sufferings as ropes and strings that would pull violently from their bodies.[45] For others, the prospect of this disease involved an unwilling acceptance of physical suffering. Doctor Falret, who wrote a book on suicide at the beginning of the nineteenth century, provided many examples. A woman who felt as though her flesh was being bitten and devoured by packs of wild dogs hanged herself with a rope tied to her bedroom ceiling; another, who was suffering from uterine cancer, poisoned herself with grains of opium and so on. In the dozens of cases studied by this expert in hypochondria, what drove the men and women to make attempts on their own lives was 'the real or imaginary pain that, having destroyed the harmony of their faculties and taken disorder to their will, led them to sacrifice the most precious of their gifts'.[46] For the Scottish doctor William Nisbet, of all the maladies to which human nature was subjected, cancer was the most formidable in its appearance: 'with a slow, but rooted grasp, it undermines the existence ... and under the torments of the most exquisite and lingering pains, as well as a state of the most loathsome putrefaction, it consigns its miserable victims to a late but long wished-for grave, after rendering them, by its ravages ... hideous spectacles of deformity'.[47]

The combination of the certainty of pain and the uncertainties of the disease came to the foreground in one of the few instances we have of a written testimony of a cancer patient during the *Ancien Régime*. Thanks to the details provided by Françoise Bertaut de Motteville, the sober biographer of Anne of Austria, we have a vivid description of the royal manner of coping with cancer in the mid-seventeenth century.[48] Though the first symptoms of her disease appeared in May 1664, it was only in November that the mother of Louis XIV began to feel significant pains in her breast and only in December that the name 'cancer' was used for the first time. Even if she began to receive external treatment with arsenic paste, no one seemed determined to make a definitive diagnosis. This was very often expressed only in the conditional tense: 'If God permits that I will be afflicted with this terrible evil that threatens me', she claimed, 'my suffering will certainly be for my own good.'[49] On 24 December, her illness was declared 'very big and incurable'. According to Motteville, Anne became aware of her condition not by looking at her own body, but at the countenance and tears of the others. Like many of her contemporaries, the former queen shared the conviction that cancer preyed especially on women and, more specifically, on women's breasts, traditionally considered a representation of charity, but also a source of vanity. This was still the opinion of the massive *Dictionary of Medicine* edited by Panckoucke between 1812 and 1860. This was also the estimation of Gilles le Vacher, a doctor from Besançon who wrote in 1740 a whole treatise on breast cancer. Immersed in a religious matrix in which every single pain had to be understood and measured in terms of imitation or retribution, Anne was convinced that her sufferings would turn her into 'an admirable model of patience and piety'.[50] After having been declared incurable, she explained to the king that God had chosen this way for the penances of her previous sins and that she was happy and prepared to die. She was then 63.

Despite her fortitude, the king made a call to the most famous physicians and surgeons of Paris. From the time of her first diagnosis with the disease until the moment of her death in January 1666, she changed doctors five times. Her first option was Antoine Vallot, first surgeon of the court, a man who, having some knowledge of chemistry, seemed to be able to provide some specific remedies. He had some ideas about the use of arsenic paste, but expressed them with so little conviction that Anne decided soon after to approach Pierre Seguin, a royal physician *à la mode*, from the Faculty of Medicine of Paris, whose therapeutic method in this, as in all other cases, was nothing other than bloodletting. His non-interventionist approach obliged Anne to abandon herself to what Motteville called 'the passions of men'. She began to listen to different opinions about doctors and was willing to receive advice about some other treatments.

The trade-offs between the patient's understanding of her own condition and the public opinions and perceptions were very clear in this case. Anne of Austria had to struggle with the symptoms of her disease, whose value and intensity were measured against a crucible of religious values and social expectations. Madame de Motteville, the king, the king's doctors and surgeons, and the other courtiers who surrounded Anne all had their own beliefs regarding the best remedies to be sought for or the best attentions to be employed: 'there were many other people who venture to suggest that they had fine remedies with which the queen could be cured', wrote Motteville.[51] Even the king commissioned different experiments and worked in close cooperation with diverse physicians.[52] In the middle of this turmoil, the borders between serious and empirical medicine, between 'rational' expectations and 'irrational' hopes became very fuzzy. The election of the appropriate remedy depended almost entirely on what someone else had heard about previous performances. Furthermore, even if the disease was thought to be fatal, Anne approached the new remedies with expectations of recovery. Despite her religious determination to endure the rodent pains of cancer and to undergo the violent pains of caustics, she hoped that she might eventually recover her health. There was a clear correlation between the way in which she coped with her pains and her secret faith in a full recovery.

Dismissing Vallot and his arsenic paste, Anne followed the recommendations of some courtiers, including Seguin himself, who suggested the secret remedies of Gendron Claude-Deshais, a village priest who, despite having no formal training in medicine, was willing to provide a cure. This was a rather unsurprising effect of the social and moral economy of hope. The business of cancer involved *dames de charité*, quacks, herbalists, chemists, apothecaries, surgeons and physicians of different kinds and conditions. The progressive failure of one treatment led, as a matter of natural course, to the search for another, including the remedies of those who, 'without title or authorisation, sell drugs and prescribe them as secrets to cure one or more diseases'.[53] In the case of Gendron, the expectations were so high that Anne herself approached him with some hopes of recovery. Though Madame de Motteville acknowledged that the remedy provided by this priest, probably a caustic, produced great pains, she also described Anne's determination to get used to these sufferings.[54] She could hardly sleep, but she did not complain; she did not show any kind of sorrow or self-commiseration. According to her biographer, those around her knew of her disease more through the sober tranquillity of her countenance than through her complaints and laments. One could only notice her condition in her eyes and in the colour of her face, explained Motteville. For Anne, as for her confidant, to whom she could speak in her Spanish mother tongue, the ordeal of her illness had come to fulfil the void of her old vanities; her sufferings had come to bleed her past sentiments of arrogance.[55] On the one hand, the memories of her old weaknesses bolstered her will to suffer; on the other hand, her sufferings were endured mainly through the acceptance and repentance of her past vanities.[56] It was not only physical suffering that bothered her mind or troubled her spirits. When she had to undress in front of everyone and expose her cancerous breast, the sad contemplation of the old instrument of her vanity produced in her no more than 'holy horror and holy indignation'.[57] Neither was pain, physical pain, the only torment of her illness. The stench that came out of her wounds was so intense that she found this suffering almost as intolerable as the other: 'the ulcerating tumours discharged so much black bile and repulsive odours that when her bandages needed changing, servants covered her nose with heavily perfumed handkerchiefs and sprinkled perfumes in the air in a vain attempt to subdue the smell'.[58] Later in the eighteenth century, Jean Astruc considered the odour of open cancer not just pestilent, but also cadaverous.[59]

In May 1665, Anne had her first serious crisis, in spite of which she remained peaceful and firm, without any exterior sign to indicate that she had any trouble in her soul.[60] The tears of the king made a strong contrast to her almost lifeless attitude and countenance; she did not show any external emotion or shed a single tear. Her constancy was bigger than her malady, wrote Motteville.[61] Only from time to time did she seem to lose her temper: 'Why do you want me to live longer? Can't you see that my life will be no more than a continuous suffering?' she cried out one day to her maid, to which Motteville replied: 'Yes, Madame, you will live on to suffer, to glorify God in your sufferings and to give pleasure to all of us.'[62] Anne did not reply. She addressed her eyes to the Heavens and, putting her hands together, she seemed to occupy herself with God and to let herself live according to His Holy Will. Though the mother of Louis XIV was always inclined to accept her fate with pious devotion, her particular *imitatio Christi* was soon overwhelmed. Her pains, wrote Motteville, were so extreme and excessive that some nights she was on the verge of desperation.

Her condition was first treated with compresses saturated in the juice of hemlock. As the disease became more threatening, her counsellors advised more innovative approaches. When her tumour began to open, Pierre Aillot, father of Jean-Baptiste Aillot, offered a different cure.[63] Unlike Gendron, he was a professional doctor, esteemed in his own country and credited among the French schools. Besides, the remedies of Gendron did not seem to have worked at all. Once Aillot was called for and despite his initial reluctance to treat the patient, for he considered it was already too late, both Motteville and Anne's attitudes were coloured with new hopes. On the other hand, however, the remedies of Aillot proved to be a new torment. Aillot 'mortified her flesh and then cut it into slides with a razor', an intervention that took place in the presence of the whole royal family, the physicians, the surgeons and all other personnel. According to Motteville, Anne endured these interventions with a notorious arrest of spirits.[64] She showed her flesh being cut with patience and sweetness and she always claimed that God had condemned her to see her flesh putrefied and rotten while she was still alive.[65] For Motteville, this was just another proof of her pious character, of her firm determination not to have her will broken. Only once did she complain loudly about her pains, and only once was she about to despair. The Countess of Flex, who was present, reminded her that it was necessary to suffer as Christ had Himself suffered on the holy cross.[66]

A few weeks before she died, Anne was also willing to call a fifth physician from Milan. This could be done through the intervention of the Spanish ambassador, who would eventually write to Italy in order to obtain news about this new remedy. At this stage, however, she was not really persuaded to leave Aillot. On the contrary, she seemed only to have a full determination to suffer and to endure the last stages of her disease. It was the desperation of her doctors rather than her own despair that played the greater role in this last attempt to find a cure. Her physicians at court had failed in every single treatment, in every single prescription. It was they who attempted to convince her that there might be some hope if this unnamed Italian were called for. On 9 January 1666, the Italian began to apply his remedies, but this only hastened her death. She passed away on 20 January.

As the story of Anne of Austria suggests, the relationship between pain and cancer touches upon three different aspects. First of all, as in many other remedies of the time, there was a great amount of iatrogenic pain, produced as a consequence of medical treatment. Second, there was the suffering produced by surgical interventions whenever this was practised. Finally, there was also the pain of cancer itself. In the case of Anne of Austria, the manner of dealing with these three forms of cancer pain was a model of imitation, one of the cultural forms in which pain acquires collective and cultural meaning.[67] She suffered while she read *The Imitation of Christ*, a devotional book. Her own understanding of the disease was framed within a religious context. She had in fact first encountered cancer when she observed how some nuns at Val-de-Grâce had died, rotten (*'toutes pourries'*), of this disease. From that moment onwards, she had been deeply frightened and horrified by this malady, which seemed so dreadful to her imagination. But the story of Anne of Austria also points to a different direction. The proliferation and use of specific remedies, either of a 'curative' or a 'palliative' nature, was always embedded with expectations based more on emotional discharges than on sober judgements. The logic of imitation, the understanding of the disease as a form of retribution but also as a way of penance and salvation, configured the way in which pain was felt and understood. The manner in which Anne was prepared to cope with the lingering pains of the disease or with the gnawing pains of its treatments cannot possibly be understood without this religious framework that turned physical suffering into an opportunity for repentance and salvation. This religious form rested, however, on a wider emotional correlation between the fear caused by the understanding of the disease and the inescapable hope for a cure. While not everyone shared Anne's interpretation of her symptoms as the result of her past vanities or her reconceptualising of the disease as an opportunity for becoming a model of piety and compassion, all those around her took part in a social economy of hope in which new healers and new remedies were always initially praised and only later dismissed.

Imaginary remedies

In the history of cancer, treatment is at the same time a need and a failure. For many authors, cancer, or at least certain types of tumours, could be cured, as their clinical histories came to prove. These hopes on the curability of the disease fed the expectations of patients and bolstered the proliferation of remedies. 'You will know how this disease is not incurable and how a remedy like the one I am about to describe is the only one able to cure it', wrote Jean Baptiste Delespine in 1772.[68] From the point of view of medical practitioners, curability depended on a set of well-defined factors, including, first of all, the ability to determine a correct observational diagnosis. Sir Astley Cooper, for example, began his *Illustrations of the Diseases of the Breast* with a clear recommendation to distinguish curable from incurable diseases:[69]

> I have scarcely witnessed a stronger expression of delight than that which has illuminated the features of a female, perhaps the mother of a large family dependent upon her protection, education and support, who, upon consulting a Surgeon for some tumour in her bosom, and expecting to hear from him a confirmation of the sentence she had pronounced upon herself, receives, on the contrary, an assurance that her apprehensions are unfounded. Pale and trembling, she enters the Surgeon's apartment, and baring her boson, faintly articulates: Sir, I am come to consult you for a cancer in my breast; and when, after a careful examination, the Surgeon states, he has the pleasure of assuring her that the disease is not cancerous, that is has not the character of malignancy, that it is not dangerous, and will not require an operation, the sudden transition from apprehension to joy brightens her countenance with the smile of gratitude, and the happiness of the moment can hardly be exceeded, when she returns with delighted affection to the family, from which she had previously considered herself destined soon to be separated by death, with the alternative only of being saved by a dubious and painful operation.[70]

In a great number of treatises and dissertations on cancer written in the eighteenth century, medical histories appealed to the distinction between the curable and the incurable, the acute and the chronic, the eradicative and the palliative. Different treatments depended entirely on the fluctuations of both sets of distinctions. The *Dictionnaire médecinal portatif*, published in 1763, considered that 'adherent cancer' could only be cured through surgery, while 'hidden cancer' would be treated with the same remedies employed in the case of haemorrhoids. In this latter case, their function was to liquefy the blood and dissolve its coagulations.[71] Whether the treatment was thought to be eradicative or palliative, the presentation of therapies involved a certain ambition of truth. The treatise of Gamet included not just the observations made on the effects of extract of hemlock on cancerous diseases, but also the minutes and certificates provided by a notary in Paris.[72] This is not surprising, since this remedy had been seriously called into question by many other doctors and practitioners. As in the case of the green lizards from Guatemala, the treatment went through different negotiations as a result of their alleged successes and their many incontrovertible failures.

Initially proposed by Storch in Vienna, some surgeons and physicians attempted to replicate the results that this doctor ascribed to the use of hemlock. For many, like Récamier, the drug was efficient only when combined with a well-designed diet. Others, like Petit, understood that failures could derive from the different characteristics of the substances employed, which had necessarily to be of the same quality and composition used in the first, successful applications.[73]

For Jean Astruc, there were three different scenarios regarding cancer treatment. First of all, there were those cases in which tumours could be operated upon, but patients did not give their consent. Second, there were also those situations in which tumours could be surgically removed and patients accepted the risk. Finally, there were many instances in which tumours could not be operated upon, irrespective of the patients' points of view. In these last cases, one could only attempt to moderate the patient's pain by means of palliative remedies.[74] In the case of Anne of Austria, the possibility of surgery was never mentioned. For her, as for many other patients, surgical interventions were the last resort. Before the late 1840s and the introduction of chemical anaesthesia, the surgical removal of tumours was a true ordeal, often compared to the martyrdom of St Agatha. For some authors, like Canquoin, the mere idea of this intervention produced an insurmountable repugnance;[75] for others, like Guillaume de Houppeville, since the surgical removal of the tumour was the only successful remedy, the pain of the intervention was easily faced and endured.[76] In either case, the personal and social evaluation of chronic cancer pain rested on a cultural matrix formed by medical criteria, practices, experiences, values and emotions. Even in those cases where cancer was declared incurable, there were plenty of remedies and treatments from which to choose. There were also a great variety of medical practices, from the non-interventionist to the most aggressive.

In one of the many theses related to the natural history of pain presented in the medical faculties of France at the beginning of the nineteenth century, cancer pain was placed under the rubric: *'chronic pain'*, a kind of pain that only ends when the patient dies.[77] For the early nineteenth-century military surgeon Garnier, however, palliative remedies were to be employed after a firm conclusion was reached on the incurability of the ailment.[78] This palliative treatment was intended to delay as much as possible the progress of the disease and to calm the symptoms of patients.[79] Garnier suggested a vegetarian diet, with abundant milk, quinine, laxatives and purgatives, and a restful and peaceful life. Opium had to be applied externally at the beginning, and only internally when pains were becoming intolerable. For Vacher, palliative treatment was to include frequent bloodletting and baths, a well-balanced diet and anodynes and narcotics. Topical remedies were to be avoided, he claimed, except in those cases in which they were required to satisfy the expectations of patients, who had great confidence in their curative powers.[80]

From our own twenty-first-century perspective, both palliative and eradicative treatments were mostly inefficient. But as the position of Garnier, Vacher and many others suggests, medical practitioners were often aware that the hopes and expectation of patients had to be satisfied even through the application of inefficient, imaginary or highly painful remedies. For Gaubius, for example, 'by the use of remedies without any effect, physicians will re-establish the tranquillity of patients, who always prefer to be piously mistaken than to be abandoned to a deadly prognosis'.[81] As in the case of Anne of Austria, both eradicative and palliative treatments came to fulfil not only medical requirements but also social hopes and personal expectations, including the hopes and expectations of physicians and surgeons. From the point of view of patients, everyone seemed to be convinced that it was much better to provide a useless remedy than not to do anything at all. Le François, in his *Projet de réformation de la médecine* from 1764, considered that doctors, when there was no remedy available, may prescribe other substances of little effect, so they will not leave their patients in that state of uneasiness that they reach when no single remedy is prescribed.[82] The same application of imaginary remedies can be found in some of the essays specifically related to pain written during the early nineteenth century. For Sarazin, for example, the sensibility of suffering diminished or even disappeared whenever the imagination was profoundly fixed on other subjects.[83] Testimony regarding the potential efficacy of painkillers is put into perspective when we read that, for Jean Astruc, for example, slices of all kinds of meats, including those of calf, hen, chicken or pigeon, worked as anodynes. To avoid the pains of cancer, he recommended the use of foils of whipped lead, either on their own or impregnated with mercury. Though he also mentioned the use of opium, especially to fight the most vivid and lacerating pains, he recommended it only in the form of poultices of tinctures of laudanum applied to the open cancer. This is reminiscent of the words of H.K. Beecher, who claimed in 1955 that: 'It does not matter in the least what the placebo is made of or how much is used so long as it is not detected as a placebo by the subject or the observer.'[84]

The green lizards of Guatemala were not available at the time of Anne of Austria. Neither was this remedy ever considered 'imaginary'.

The conviction shown by many authors of the beneficial effects of the intake of this reptile flesh is indicative of many other similar stories related to other substances of different origins. The proliferation of specific remedies for the cure of cancer and the treatment of its most repugnant symptoms should be understood within a social logic of hope and resistance. Even if the name of the disease was rather hidden and the condition considered 'incurable', this social apprehension did not imply the acceptance of an irremediable fate. On the contrary, the medical and paramedical treatises on cancer published in Europe during the early modern period are filled up with stories of patients who recovered their health or who diminished their sufferings by the intervention of general or specific remedies. We may very well frame those stories within the general history of the placebo effect. We may also call into question the credibility of those testimonies. We may also wonder to what extent the pains of the disease, and the medical histories of its treatment, were embedded with cultural values and social expectations. What I have called here 'the moral economy of hope', the cultural form in which pain and fear are counterbalanced by promises and expectations, deserves, in my view, a more detailed analytical look and a more refined historical understanding.

Notes

1. Research for this chapter has been made possible through two research grants provided by the Spanish Ministry of Economy (FFI2010-20876) and by a *séjour de recherché*, funded by both the Ministry of Culture and the Foundation de la Maison des Sciences de l'Homme. I would like to thank my colleagues at the Centre Alexandre Koyré, the librarians of the Bibliothèque Interuniversitire de Histoire de la santé of the University Paris V, René Descartes, in Paris, and of the Wellcome Library for the History of Medicine in London for their kind support.
2. The complete title was *Específico nuevamente descubierto en el Reyno de Goatemala, para la curación radical del horrible mal de cancro y otros más frecuentes* (Cadiz, 1783). On the life of this medical doctor, see José Aznar López, *El Dr. Don José de Flores, una vida al servicio de la ciencia* (Guatemala: Editorial Universitaria, 1960); and José Luis Maldonado Polo, *Las huellas de la razón. La expedición científica de Centroamérica* (Madrid: CSIS, Estudios de Historia de la Ciencia, 2001), 192–202; Jorge Cañizares Esquerra, *Cómo escribir la historia del nuevo mundo. Historiografías, epidemiologías e identidades en el mundo del Atlántico del siglo XVIII* (Mexico: FCE, 2007), 480–3. On the debates related to the lizards, see Miruna Achim, *Lagartijas medicinales: Remedios americanos y debates científicos en la ilustración* (Mexico: Conaculta/UAM Cuajimalpa, 2008).
3. On the frequency of cancer at the end of the Ancien Régime, see J. Meyer: 'Une enquête de l'Académie de médecine sur les épidémies (1774–1794)',

Annales. Économie, Sociétés, Civilisations, 21(4) (1966): 729–48; J.P. Peter: 'Une enquête de la Société royale de médecine (1774–1794): malades et maladies à la fin du xviii siècle', *Annales. Économie, Sociétés, Civilisations*, 22(4) (1967): 711–51. See also J.P. Desaive, J.P. Goubert, E. Ley Roy Ladurie, J. Meyer, O. Muller and J.P. Peter, *Médecines, climat et épidémies à la fin du xviiième siècle* (Paris: Éditions de l'École des hautes études en sciences sociales, 1972).

4. S. Tanchou, *Recherches sur le traitement médical des tumeurs cancéreuses du sein* (Paris: Germer Baillière, 1844), 254, quoted by Jacques Rouëssé, *Une histoire du cancer du sein en Occident: enseignements et réflexions* (Paris: Springer, 2011), xxiii.

5. René de Bovis: 'L'augmentation de fréquence du cancer, sa prédominance dans les villes et sa prédilection pour le sexe féminin sont-elles réelles ou apparentes?', *La semaine médicale*, 36 (1902): 297–302.

6. Bertrand Bécane, *Observations sur les effets du virus cancéreux, divisées en cinq questions* (Toulouse: Imprimerie de J.F. Desclassan, 1778), 22. For a general history of cancer, see Siddhartha Mukherjee, *The Emperor of All Maladies: A Biography of Cancer* (New York: Scribner, 2010); Rouëssé, *Une histoire du cancer*. A book that links the history of cancer with the history of fear is Pierre Darmon, *Les cellules folles* (Paris: Plon, 1993). More recent contributions include Didier Foucault (ed.), *Lutter contre le cancer (1740–1960)* (Toulouse: Privat, 2012). In Spain, Fanny H. Brotons: 'Por una historia cultural de la enfermedad: la experiencia del cáncer en la España de la segunda mitad del siglo XIX' (Master's thesis. University Carlos III, Madrid, 2013).

7. See Miruna Achim, *Lagartijas medicinales*, 10.

8. Flores, *Específico nuevamente descubierto*, 5.

9. Howard Spiro, 'Clinical Reflections on the Placebo Phenomenon', in Anne Harrington (ed.), *The Placebo Effect: An Interdisciplinary Exploration* (Cambridge, MA: Harvard University Press, 1997), 37–55 at 42.

10. See Darmon, *Les cellules folles*, 7–12. According to Darmon, the link between cancer and fear is a twentieth-century phenomenon. The third part of his massive book deals with this connection: 'La grande peur (1920–1990)'.

11. Martinet, *Essai [...] sur l'usage des lezards* (Palermo: Gastelier, 1786), 39.

12. For the history of the 'placebo effect', see Harrington, *Placebo Effect*; S.W. Jackson, *Care of the Psyche: A History of Psychological Healing* (New Haven: Yale University Press, 1999). For a more recent approach, see Daniel Fecteau, *L'effet placebo. Le pouvoir de guérir* (Quebec: Les éditions de l'homme, 2005).

13. See Jason Szabo, *Incurable and Intolerable: Chronic Disease and Slow Death in Nineteenth Century France* (New Brunswick: Rutgers University Press, 2009).

14. J.M. Gamet, *Traité des affections cancéreuses, pour servir de suite à la théorie nouvelle sur les maladies du même genre* (Lyon, 1777), 43–4.

15. H. Boerhaave, *Praelectiones academicae in proprias instituciones rei medicae*, 6 vols. (Göttingen: A. Vanderhoeck, 1794), vi, 278–9, cited by Daniel De Moulin, *A Short History of Breast Cancer* (Dordrecht: Martinus Nijhoff Publishers, 1983), 34.

16. Jean Astruc, *Traité de maladies de femmes*, 6 vols. (Paris: P.G. Cavelier, 1770), vol. 5, 314–39. Quoted by De Moulin, *Short History of Breast Cancer*, 34.

17. L.J.M. Robert, *L'art de prévenir le cancer au sein chez les femmes qui touchent à l'âge critique* (Paris: Crochard; Marseille: Jean Mossy, 1812), cited by Dominique Gros, *Cancer du sein. Entre raison et sentiments* (Paris, Springer-Verlag, 2009), 50.

18. Alexandre Canquoin, *Traitement du cancer. Exposé complet de la méthode du docteur Canquoin excluant toute opération par l'instrument tranchant* (Paris: Alexandre Canquoin, L'abbé Bechet, 1838), 8 and 11.

19. J.A. Récamier, *Recherches sur le traitement du cancer par la compression simple ou combinée, et sur la histoire générale de la même maladie, suives de notes* (Paris, 1829), 10–11.

20. Bernard Perylhe, *A Dissertation on Cancerous Diseases* (London: J. Wilkins, 1777), 44.

21. Guillaume de Houppeville, *La guérison du cancer au sein* (Rouen: Guillaume Benhort, 1693), 67–8.

22. René Ledoux-Lebard, *La Lutte Contre Le Cancer* (Paris: Masson et Cie, 1906).

23. Perylhe, *A Dissertation*, 87.

24. Canquoin, *Traitement du cancer*, 29.

25. M. Martinet, *Observations médico-chimiques sur la cancer* (Paris: De l'imprimerie de monsieur, 1783).

26. Récamier, *Recherches sur le traitement du cancer*.

27. John Brisbane, *Select Cases in the Practice of Medicine* (London: T. Cadell, 1773).

28. J. Weldon Fell, *Treatise on Cancer, and its Treatment* (London: John Churchill, 1857), 56ff.

29. Daniel De Moulin, *Short History of Breast Cancer*, 50.

30. Guillaume René Le Febure, *Remède éprouve pour guérir radicalement le cancer occulte & manifeste ou ulcéré* (Paris: Michel Lambert, 1775).

31. Canquoin, *Traitement du cancer*, 46.

32. Jean Astruc, *Traité des tumeurs et des ulcères* (Paris: Guillaume Cavelier, 1759), vol. 2, 62–3.

33. De Moulin, *Short History of Breast Cancer*, 43.

34. See, for example, the openings remarks by Canquoin, *Traitement du cancer*, 1: 'De toutes les maladies qui affligent l'espèce humaine, il n'en est guère de plus redoutable que le cancer, comme il n'en pas de plus décourageante pour l'homme de l'art.'

35. On the contending sides of the prescription of medicines, see Pascale Gramain-Kibleur, 'Le monde du médicament à l'aube de l'ère industrielle. Les enjeux de la prescription médicamenteuse de la fin du xvii au début du xix siècle' (thesis, University of Paris, 1999).

36. Walter Hayle Walshe, *Nature and Treatment of Cancer* (London: Taylor and Walton, 1846), 120–1.

37. J.B. Aillot, *Traité du cancer, où l'on explique sa nature, & où l'on propose les moyens les plus sûrs pour le guerir methodiquement* (París: chez François Muguet, 1698), 19.

38. J. Burrows, *Nouvel essai de médecine pratique sur les cancers* (París: Dessain, 1767), 5.

39. Astruc, *Traité des tumeurs*, ii, 35–6; Jean-Henri Boniol, *Dissertation sur le cancer aux mamelles. Thèse présentée et soutenue à l'École de Médecine de Montpellier, le huit Messidor, an six de la République* (Montpellier: Jean Martel, 1798), 30.

40. Burrows, *Nouvel essai*, 42.

41. J.A. Helvetius, *Lettre à Monsieur Régis, sur la nature et la guérison du cancer* (Paris: Jean Cusson, 1691), 5.

42. See Houppeville, *La guérison du cancer au sein*. See also D. Gros, *Cancer du sein. Entre raison et sentiments* (Paris: Springer-Verlag, 2009), 81.

43. Gilles le Vacher, *Dissertation sur le cancer des mamelles* (Besancon: Jean-Baptiste Charmet, 1740), 10.
44. Perylhe, *A Dissertation*, 48, 93.
45. Helvetius, *Lettre à Monsieur Régis*, 4–5.
46. Jean Pierre Falret, *De l'hypochondrie et du suicide. Considérations sur les causes, sur le siège et le traitement de ces maladies, sur les moyens d'en arrêter les progresses et d'en prévenir le développement* (Paris: Croullebois, 1822), 157.
47. William Nisbet, *Inquiry into the History, Nature, Causes and Different Modes of Treatment in the Cure of Scrophula and Cancer* (Edinburgh: Alex Chapman & Co., 1795), 4.
48. Françoise Bertaut de Motteville, *Mémoires de Mme de Motteville sur Anne d'Autriche et sa Cour* (Paris: Charpentier, 1855), vol. 4. See also Ruth Kleinman: 'Facing Cancer in the Seventeenth Century: The Last Illness of Anne of Austria, 1664–1666', *Advances in Thanatology*, 4 (1977): 37–55.
49. Motteville, *Mémoires*, vol. 4, 365.
50. *Ibid.*, 365.
51. *Ibid.*, 367–8.
52. *Ibid.*, 386.
53. Jean Emmanuel Gilibert, *L'anarchie médicinale, ou la medecine considérée comme nuisible a la société*, vol. 1 (Neuchatel, 1772), 254.
54. Motteville, *Mémoires*, vol. 4, 367.
55. *Ibid.*, 368.
56. *Ibid.*, 369.
57. *Ibid.*, 405.
58. *Ibid.*, 429; trans. and quoted by James S. Olson, *Bathsheba's Breast: Women, Cancer and History* (Baltimore: Johns Hopkins University Press, 2002), 25.
59. Astruc, *Traité des tumeurs*, vol. 2, 46.
60. Motteville, *Mémoires*, vol. 4, 378.
61. *Ibid.*, 382.
62. *Ibid.*, 384.
63. Aillot, *Traité du cancer*.
64. Motteville, *Mémoires*, vol. 4, 405.
65. *Ibid.*
66. *Ibid.*, 404.
67. See Javier Moscoso, *Pain: A Cultural History* (Basingstoke: Palgrave Macmillan, 2012), Chapter 2, 'Imitation'.
68. Jean-Baptiste Deslespine, *Idée générale des causes de cancer et de la manière de le guerir* (Paris: Delespine, 1708), 4.
69. Astley Cooper, *Illustrations of the Diseases of the Breast* (London: Longman, 1829), Part 1, 2.
70. *Ibid.*, 3–4.
71. Jean Guyot, *Dictionnaire médicinal portatif, contenant un méthode sûre pour connaitre & guérir les Maladies critiques & chroniques par des remèdes simples & proportionnées à la connaissance de tout le monde* (Paris: d'Houry, 1763), 37.
72. Gamet, *Traité des affections cancéreuses*, 50.
73. Canquoin, *Traitement du cancer*, 23.
74. Astruc, *Traité des tumeurs*, vol. 2, 54.
75. Canquoin, *Traitement du cancer*, 25.
76. Houppeville, *La guérison du cancer au sein*, 35.

77. J. Sarrazin, *Essai sur la douleur, présenté et soutenu à l'École de Médecine de Paris* (Paris: Imprimerie de Didot Jeune, 1805), 24–5.

78. G.J. Garnier, *Dissertation sur le cancer* (Paris: De l'imprimerie de Delance at Lesueur, an xii. [1803]), 16.

79. *Ibid.*, 16.

80. Gilles le Vacher, *Dissertation sur le cancer des mamelles* (Besancon: Jean-Baptiste Charmet, 1740), 101.

81. H.B. Gaubius, *L'art de dresser les formules de médecine* (Paris: Dessaint & Saillant, 1749), 5–6, quoted by Rouësse, *Une histoire du cancer*, 136.

82. Gramain-Kibleur, 'Le monde du médicament', 29.

83. Sarrazin, *Essai sur la douleur*, 28.

84. H.K. Beecher: 'The Powerful Placebo', *Journal of the American Medical Association*, 159(17) (1955): 1602–6 at 1602.

3
The Perception of Pain in Late Imperial China

Paolo Santangelo

This chapter can only offer some hints on the cultural background and the perception of pain in Late Imperial China, and will limit its scope to an analysis of the main terms used for physical suffering and pain, presenting some examples in order to better understand how pain is represented, described and located in the collective imagination.[1] It will clearly appear that physical pain was understood not only as a physical phenomenon, but also as involving the whole body-mind-heart system. Moral suffering was also related to the body. Cognitive factors involved the roles of values in the emotional process and in the evaluation and perception of bodily sensations. This meant that psychological and cultural factors were fundamental even in physical perceptions. Thus, the specific characters of a culture – intellectual means, categories, symbols, imagery and so on – concurred with the representation and the consciousness of pain.

Pleasures of life, as well as suffering, experiences of pain and illness, and even death are objects of different representations in different cultures. Understanding sensations means understanding the most essential reactions of our beings, reflecting on human perceptions of pleasure and pain, and re-creating new images of the self. What I intend to take into consideration is the idea of a continuous interaction between emotions and sensations, which occupied writers and thinkers of different cultures.

All physical perceptions are not just 'objective' or 'mechanical' reactions of the human body or its parts; the perception of bodily sensations is deeply influenced by the cultural context.[2] This phenomenon is evident in many physical reactions, from sexual arousal to other physical perceptions, and sensibility can be more or less activated according to personal disposition, emotional mood, habits and cultural tendencies.

From our own experience we learn that even a sensation like pain has different meanings in different contexts and cultures.[3]

Thus, it is reasonable to assume that physical states have a psychological aspect, while psychical moods are both influenced by health, hormonal conditions and physical sensations, and have an influence on the body's conditions.[4] This is valid for all sensations, including pain and physical suffering: aches, chills and fever, nausea, dizziness, extreme fatigue and insomnia, bodily degradation and physical restriction in illness, ageing and agonizing pain are all conditions that are endured by the subject and cannot be shared or easily communicated. But they are often connected to and can give rise to feelings of depression and sadness. In any case, the representation of these states in fiction and reports resort to different rhetorical devices, and, moreover, evidence the need of sublimation or mythical explanation, the strategies people use to construct a sense of order and meaning from chaos and disintegration.[5] These seem to be universal features that can also be found in the representations of pain in Chinese culture.

This close interconnection is constantly confirmed both by the stress of psychological elements in physical states and by health conditions in psychical moods and physical sensations, or by the well-known phenomenon of *somatisation*, such as chest pain or fatigue, which may reflect specific cultural constructions of the individual, who reacts differently according to her social background.[6] This close relation is evident in an almost universal phenomenon, which is also evident in Chinese language, where terms originally used for physical reactions had their usage extended to the more symbolic meaning of the sentiments: the semantic ambiguity associates pain with the subjective-behavioural-emotional phenomenon of suffering by mingling both concepts.

The most common characters used to express such negative experience are significant: *tong* 痛, 'painful'; *ku* 苦, 'bitter', 'hardship', 'suffering', 'pain', 'cause somebody suffering', 'suffer from', 'be troubled by'; and their compounds, such as *tongku* 痛苦, 'suffering', 'anguish'. This interaction between the two phenomena of physical pain and moral pain is evident in the equivalents given by dictionaries. Even the physiological quality of pain, such as burning or stabbing, its intensity and its manifestation in time, are extended to express mental sufferings or emotional states. This semantic ambiguity metaphorically represents physical experiences through mental images. It appears under different perspectives and situations, as in the case of psychosomatic symptoms, suggesting that mental pain is felt concurrently with physical pain or is caused by it. Moreover, the perception of actual or threatened damage, the experience of unpleasantness, with the implicit evaluation of hurting and damaging, is expressed by the term *shang* 傷, 'injure', 'wound', 'fall ill from', 'damage', 'harm', but also 'sad', 'distressed'. Analogous are *shang* combinations, such as *shangbei* 傷悲, *youshang* 憂傷, *shanghuai* 傷懷 (respectively, sad-grief-distress, worried-grieved and broken-hearted), and related words from the semantic field of suffering, most notably expressions of anxiety and illness.[7]

Pain is one of the most frequently described bodily sensations: the terms used include 'ache' 作痛, 'pain' 疼痛, 'suffering sore' 吃疼, 'painful and weak four limbs' 四肢酸軟, 'be distressed/love dearly' 心疼, 'abdominal pain' 腹痛, as well as metaphors, such as 'needles' 針挑刀掏, 'knives' 刀子割, 'baking fire' 火炙 and 'hitting the head' 攔頭. Other expressions are no less strong, such as 'shiver intestines and search through lungs, roast stomach and stir up liver' 抖腸搜肺、炙胃扇肝的 or the ambiguous 'a deep pain in one's heart' 心口疼痛. The reaction is often expressed with crying and laments (噯喲, 嘆).

Even ancient Chinese medical treatises, such as the 'The Yellow Emperor's Inner Canon' *Huangdi neijing* 黃帝內經, show a clear consciousness of interaction between physical and emotional spheres, and emphasise the intimate connection between moral and physical pain, and the reciprocal influence of emotions and bodily sensations. As Angelika Messner notes, 'Emotions like joy, anger, sadness, and fear, and bodily sensations like hot, cold, fatigue and pain as presented in various *Huangdi neijing* passages are processes or temporary states (*Befindlichkeiten*)', which are clearly described in bodily terms, in contrast to earlier models of explanation where they mostly appeared as 'outer atmospheres seizing men abruptly'. Her article shows how the Western conception of 'emotion and bodily sensation', which per se reflects a dualistic separation, is different from the specifically Chinese medical conceptions, which are not congruent with this dualistic separation. Rather, they describe the dimensions of basic bodily and emotional states and movements on the basis of the *qi*-flow. In fact, 'physical appearance' (*xing* 形), 'emotions/mood/feeling' (*zhi* 志) and 'spirit/vital force' (*shen* 神) are all related to changes of *qi*-flow, both emotional changes and bodily sensations, as emotions had nothing to do with individualistic private states within one single 'place', such as a 'soul' or a 'psyche'.[8]

The inseparable connection between 'bodily' sensation and 'spiritual' emotion can also be found in philosophical productions, starting from *Mencius* and *Xunzi*. The virtue of humaneness (*ren* 仁) in *Mencius* is simultaneously sensation (sight of the suffering ox 見牛), emotion (not-bearing不忍), moral consciousness for compassion (*ceyin zhi xin* 惻隱之心) and feeling (the suffering of the other is transferred in the subject: *ce* 惻 and *yin* 隱 both mean deep and profound pain). Thus, what cannot be shared or easily communicated becomes partaken of thanks to the virtue of humaneness and compassion. Bodily pains could never be only 'bodily' or 'physical' sensations because they necessarily 'move' (*dong* 动) one's 'heart-mind'. Emotions are grounded in bodily sensations, as the human condition is embodied in a physical body that is human sensitivity.[9]

In Xunzi's writings there is also a close relation between *qing* 情, passions and sensory feelings. As Nylan notes:

> Given that the evaluative impulses are endowed at birth, the self inevitably seeks that which it accounts a satisfaction, supplying what it perceives itself to lack. A human has no 'value-free', 'neutral' response. Instead, a preliminary assessment of a particular phenomenon's value to the self disposes the person to want to secure or to shun the phenomenon. These dispositions are classified by level of intensity: if a person finds a particular phenomenon pleasing, he/she may feel a liking or preference (*hao* 好) for it, a frisson of delight (*xi* 喜), or a more lasting sense of pleasure (*le* 樂). On the other hand, if the phenomenon fails to please, she may feel dislike or distaste (*wu* 惡), a spurt of anger (*nu* 怒), or a more profound and lasting sense of pain and loss (*ai* 哀). These six basic dispositions (*qing*), endowed at birth, represent the initial, unmediated inclinations to act.[10]

And again, several centuries later, a Ming thinker, Lü Kun 呂坤 (1536–1618), stressed another aspect, noticing that when one is oppressed by pain, he may feel depressed by this feeling: 'We do notice and pay attention even to our minimal sufferings' (LK 172).[11]

Among the various sources that can give us information on the representation, evaluation and perception of pain, the richest are certainly literary works that present everyday life and common perceptions of feelings. Although they are not strictly speaking historical sources, they offer many data that cannot be found in other materials, from official chronicles to individual reports. Chronicles and official histories rarely take into consideration individual physical suffering. A rare example is a note composition between a literary work and an historical account, which witnesses the interference between the body and the heart-mind. For instance, the 'Record on the Ten Day [Massacre] at Yangzhou' (*Yangzhou shiri ji* 揚州十日記) of 1645 deals with pain and suffering in the course of the traumas people experienced during the fall of the Ming Dynasty and the conquest of the Empire by the Manchus. In this text, physical pain due to cruel physical injury, as well as emotional despair (*tong* 痛 and *shang* 傷), are textualised throughout in terms of visceral processes and changes.[12] In the tales, various kinds of pains are described (serious, burning, stabbing), but in many cases the unbearable pain is prevailing: 痛不可忍, 不勝痛, 痛不可當, 痛不可耐 (all meaning 'unbearable pain').

Influenced by medical points of view, the well-known writer and publisher Feng Menglong 馮夢龍 (1574–1646), with a focus on women's suffering, leads the readers' attention to her five viscera and four limbs. In one of his stories he describes a woman who collapses on the floor after having heard of her child's death. The author comments: 'If you don't know what her five *zang* organs are about, first see how she lies there with her four limbs not able to rise [any more] 不知五臟如何, 先見四肢不舉.'[13] The pain associated with illness of the inner organs or the skin inspired metaphors and metonymies of psychic pain. Evidently, observing these corporeal conditions was supposed to reveal an idea about her painful suffering:

> Prefect Wang Mengting suffered from an ache at his waist and so he asked the Daoist for help. The Daoist said: 'I will come to cure you when the weather improves.' On the first fine day, Lü gathered sunshine with his fingers and then rubbed his hand over Wang's waist. Wang felt heat travel through his five internal organs and he was cured. (ZBY 09:182, Lü Dao Ren Qu Long 呂道人驅龍)

Thus, the notions 'five viscera' (*wu zang* 五臟) and 'four limbs' (*si zhi* 四肢) are more than just a rhetorical play on words. The five viscera may become a metaphor for the heart, but the suffering-body relation, as in the 'Dream of the Red Mansion' (*Honglou meng* 紅樓夢), is retained: 'I've broken my heart, yet you're still weeping' (我的五臟都碎了, 你還只是哭) (HLM 30:454).[14] The perception of pain is clearly influenced by medical texts when it is related to the imbalance of the vital body energies. Thus, the primary cause of pain is sickness and poor health, apparently limited to the bodily and to the physical: 'The predominance of the wood element in the liver over the earth element in the spleen causes loss of appetite, general lassitude and soreness of the limbs' (HLM 10:170).[15] In the analysis of sources, we can envisage the pure physical sensations caused by external impact or inner health conditions, and sensation caused by emotions. But even in the former case, the connections between moods and states of mind may become evident. In an episode of the 'Dream of the Red Mansion', the main character Baoyu has been beaten by his father: 'Although Baoyu was lying on bed as quietly as he could, his buttocks were painful as if pricked by needles, cut by knives or scorched by fire. Thus, the slightest movement wrung a groan from him – *aiyo!*' (HLM 34:499). While here only the physical sensation is mentioned, in the following examples the perception of the relief comes from calling the name of Baoyu's beloved girls:

Therefore more than once, his father thrashed him within an inch of his life, but still that didn't change him. Whenever the pain became too much for him, he would start yelling, 'Sister', 'Little sister' … 'Once, [he said] when pressed by pain, I called the girls in the hope of alleviating the soreness, without knowing the result. And because after I had called them, I really felt the pain diminish somewhat, now that I have obtained this secret spell, whenever I feel the extremity of pain, at once I call the girls continuously.' (HLM 2:30)

In the following example of psychosomatic reactions, it is sorrow that manifests itself as pain:

Lady Wang could not refrain from weeping in secret, as she mourned her daughter, grieved for her brother and worried over Baoyu; and now this third misfortune in swift succession was more than she could bear – she came down with colic … At these words, Lady Wang, in a pang of sadness, felt a deep pain in her heart, and was unable to remain seated… (HLM 96:1431)

The psychosomatic effects of sorrow are manifested in real pains felt inside the body. The following passages clearly express how states of mind were considered to influence sensibility, either in feeling cold or pain, or in inducing insensibility:

'I don't know what Miss Zijuan's told him, but the boy's eyes are dully staring, his hands and feet are cold; he can't speak a word, and he felt nothing when Nanny Li pinched him. He's more dead than alive! Even Nanny Li says there's no hope and is weeping and wailing there. He may be dead by now for all I know.' Nanny Li was such an experienced old nurse that Daiyu could not but believe her gloomy predictions. With a cry she threw up all the medicine she had just taken, and was racked by such dry coughing that her stomach burned and it seemed her lungs would burst. (HLM 57:869)

He was so moved that all sense of pain flew at once beyond the clouds of heaven. (HLM 34:498)

Oh, Chunxiang, take good care of your young mistress, who is hurt by spring and fears the summer. One that is troubled by sickness is most easily saddened by the fading autumn days. Miss, I will go and prepare your medicine. (MDT 18:102)

The last example, on the contrary, expresses the awareness of the influence of pain and sickness on one's mood. It is a different case when a physical pain is attributed to a physiological or mechanical cause. The etiology of pain may appear rather obvious and confirms some universal conditions related to human sufferings, but it is worth examining the main cases presented by literary sources, since they reflect the common way of thinking in that society. Physical pain may be a result of eating some unhealthy food or an excessively abundant meal, or of staying for a long time in the same position: 'You two had better not eat too much either. Crabs are delicious but not very wholesome. If you overeat, you'll have a stomach ache' (HLM 38:560); 'My heel is so achy after such a long time of standing' (ZBY 05:091, Xi Zi He Che 洗紫河車). A headache can also be caused by an excessive noise or bad music, as in the following example taken from Chapter 54 of the 'Dream of the Red Mansion', on the opera 'The Eight Worthies' (八義): 'The eight acts from *The Eight Worthies* were so noisy that they've made my head ache. Let's have something quieter' (HLM 54:826). Other external causes include excessive activity (being tired), smell or touch, like the touch of hot things, or bumping against some hard object: 'He burns hot his own hand and asks someone else if it hurts or not. Isn't that very stupid?' (HLM 35:518); '*Gudong*! with a crash, she fell against the wooden partition and bumped her head, which began to hurt' (HLM 41:618); 'Madam, be careful not to hurt your hand!' (HLM 44:657).

Women suffer both in pregnancy and in giving birth. Worth mentioning is a popular song collected by Feng Menglong, which tells how the pain is forgotten through the joy of the event rather than because of medical infusions:

She felt a bellyache and she sipped a ginger soup; at midnight she gave birth to a male baby in her private boudoir; holding him with her slender jade fingers, she looks at him under a red lantern; he takes half after me and half after you! (*Shan'ge* 1:32D)

Illness itself can also be related to psychological conditions, and in Chinese tradition the relation is very close. According to Yang Lai'er (楊萊兒), exaggerated passions are the cause of many illnesses (多情多病)[16] and maybe nobody would have questioned the secret of the eternity of Heaven because 'if Heaven had passions, it would also become quickly ill and old' (天若有情, 天亦老).[17]

Pain is the symptom of a serious illness that often leads the subject to death:

> Shen woke up from his dream with an excruciating pain in his stomach. He called his colleagues for help and told them about his experience in the dream. Three days later, he died. (ZBY 09:171, Cheng Huang Shen Xu Jiu 城隍神酗酒)

> ... but the ghost grabbed him hard and clutched his testicles. The pain was unbearable and Chen woke up with a start. His scrotum was swollen and had become as enormous as a *dou*. He began suffering from alternating spells of fever and chills. The doctor could do nothing to cure him, and within a short amount of time Chen died in the same student dormitory. (ZBY 11:215, Zhang You Hua 張又華)

Thus, illness is frequently presented as a cause of pains, as in the following examples:[18] 'Unexpectedly, in the middle of the night, he repeatedly shouted from pain in his heart, talking nonsense, as if he were being stabbed by a knife' (HLM 83:1262); 'After breakfast, the attack indeed started again, with special pains. Shao came to practice acupuncture on her and soon she recovered' (LZZY 7:892, Shao Nü 邵女); 'On the following day, Guo had a bad bellyache. His bowel movement was as green as tarnished bronze' (LZZY 7:915, Guo Xiu Cai 郭秀才); 'The patient woke up from coma, saying: "Why do I feel extremely painful all over?"' (ZBY 06:114, Shen Xing Qi 沈姓妻); 'Suddenly the daughter of Zhou got a strange illness. At first she felt her heart was painful and then her stomach and back was painful too. Finally her ears, eyes, mouth and nose were all painful. She wailed and jumped and writhed that people couldn't bear to see' (ZBY 10:185, Hou Guai 猴怪).

Love-sickness, *xiangsibing* (相思病), 'longing-disease',[19] is the special melancholy that does not let people either speak or eat (不語亦不食)[20] and sleep (以致廢寢忘餐).[21] In narrative and theatre, from the cycle of Ying Ying and of the Western Chamber to Daiyu and other characters in the 'Dream of the Red Mansion',[22] it appears under various perspectives, either in the condemnation of passions as a purely pathological phenomenon or its literary sublimation, presenting the 'bitter' aspects of love: loneliness after unreciprocated love and the suffering of separated lovers and abandoned or suspicious partners make up the majority of states of mind related to this sentiment.

Pain, finally, may be induced by many causes, by natural events or by others' violence. Most of the examples of pain come from violence, from being beaten or from a wound:

> He [Xianglian] fetched his horsewhip and gave him [Xue Pan] a few dozen strokes all over his back ... [Xue Pan] found the pain so intolerable that he could not but lament '*aiyo*' ... Xianglian tossed away the whip to pummel him with his fists. Xue Pan rolled over and over, frantically howling, 'Oh, my ribs are broken!' (HLM 47:703–4)

> He beat the scholar and spat at him. The scholar couldn't bear the pain and he begged for help to all the others in the market. (ZBY 11:204, Liu Gui Sun Feng 劉貴孫鳳)

> I was flogged twenty times ... Now my private part is so painful. (ZBY 06:117, Chang Shu Cheng Sheng 常熟程生)

> You have clamped our legs and it was too painful to bear. (ZBY 06:111, Men Jia Gui Tui 門夾鬼腿)

> I felt an unbearable pain and so I touched my head. My left ear was gone and the blood flowed down ceaselessly. (ZBY 10:184, Yu Wang Bei Tun She 禹王碑吞蛇)

And, in the Buddhist perception of rebirth, animals suffer no less than human beings, as with the horse when mistreated:

> When the master rode him, he always put a mudguard under the saddle, loosened the rein, driving gently and slowly, so it wasn't bad; but when the servants or the groom rode him, they never used padding, and kicked his belly with their heels, which was so painful that he felt the soreness penetrating deeply inside. (LZZY 1:73, San Sheng 三生)

In fact, the most brutal tortures were those suffered in the Buddhist Hells, as the above story goes on to describe:

> Finally, he got so angry that he refused to eat for three days and died. So once again, he was back in Hell. The king of Hell checked

with the life-and-death book and found that Liu had not finished serving his term. He reprimanded Liu for intentionally escaping his punishment. So he was condemned to have his skin peeled off and to become a dog in his next life. Regretful and depressed, Liu did not want to go, but a group of demons gave him a thorough beating which hurt him so much that he ran off to the countryside. He thought to himself: 'Why, it's better to die than to live like this'. So he jumped down from the cliff with indignation and fell down to the ground, unable to get up. (LZZY 1:73, San Sheng 三生)

The internalisation of such fears might have allowed an acute repentance and a sense of guilt to cause analogous sufferings, as in this surrealistic and allegoric passage from Pu Songling's tales:

Once he recalled a meritorious deed, his heart was peaceful and clear; when a bad action, he was regretful and worried, like being put in a cauldron with boiling oil. The suffering was unbearable and could not be described. He still remembered that when he was seven or eight, he once drew a bird-nest and the young birds died. Only thinking of this act, his heart was burning with a warm tide and this feeling passed after a short moment. (LZZY 3:326, Tang Gong 湯公)

Such sufferings can be compared only with hell's tortures or with stereotyped punishments inflicted by jealous wives against the maids and concubines of their families:

So Jin took a hot iron with the intention of burning the girl's face. The other servants felt indignant and uneven. Every time the girl cried for pain, all servants cried their hearts out for mercy and begged to take Shao's place, volunteering to die instead of her. Only then Jin stopped torturing her, but stabbed her with a hairpin twenty times in the chest before letting her go (LZZY 7:890, Shao Nü 邵女)

Pain was sometimes accompanied by hallucinations.[23] Spirit possessions could provoke pain and also extinguish such pain (絕無痛楚), as in the following example, where a female ghost haunts a man, pressing him to suicide in order to also become a ghost and then to marry her:

Several days later they heard a high-pitched scream upstairs. They rushed up and found scholar Wu collapsed on the floor with a knife plunged into the right side of his abdomen, some of his intestines spilling out through the wound. His throat had also been cut, severing his oesophagus. His colleagues helped him up but it was clear that he felt no pain. When the magistrate Lu arrived to investigate the situation, Wu beckoned him closer and then wrote the word 'injustice' in the air. 'What injustice?' asked Magistrate Lu. Wu then explained: 'We are quarrelsome lovers, two people destined to be a couple in spite of all the animosity. This morning the woman came again and forced me to kill myself so that we could become husband and wife in the netherworld. I asked her how I could kill myself, and she pointed to a knife on the table, saying that that was the best way. I took the knife and plunged it deep into the right side of my stomach: the pain was unbearable 痛不可忍. The woman came over to massage the wound 婦人亟以手按摩之 and said: "No, this [injury] won't do." However, I did not feel any pain where she had massaged me 所摩處遂不覺痛. I asked her what I needed to do. And she made a gesture as if she were cutting her own throat, urging me to do so. So I picked up the knife again and slashed at my throat. But this only made her stamp her feet and sigh: "This is also useless; you're just causing yourself pain 徒多痛苦耳." She massaged my throat and the pain disappeared 以手按摩之, 亦不覺痛.' (ZBY 07:138, Wu Sheng Shou Ruan 吳生手軟)

Again, popular magic beliefs and traditional medicine could hail sufferings, especially when their psychological components were dominant:

Zhou's daughter developed a mysterious illness. At first her heart hurt and then so did her stomach and her back. Finally, the pain spread to her ears, eyes, mouth and nose 始而心痛，繼而腹背痛，繼而耳目口鼻無不痛者，哀號跳擲. She wailed and jumped and writhed so much that it was unbearable to watch. Wu sent for all the doctors but none of them were able to diagnose her disease ... The monkey drew out several prods, iron needles and bamboo spikes from the girl's eyes, ears, mouth and nose. Her pain was alleviated everywhere except in her heart 猴抉其眼耳口鼻中，所出橫刺、鐵針、竹籤十餘條，女痛稍蘇，惟心痛未解 ... The monkey then stretched her hand into the girl's throat all the way down to her chest and pulled out a copper mirror, still stained with blood. The girl regained her health at once. (ZBY 10:186, Hou Guai 猴怪)

Pain might have ended after recovering from illness or the decreasing effects of offence. Medicines might have had analgesic effects, as too might have miraculous pills: 'After saying that, he took out the pill and put it into Ren's mouth. Ren's ache ceased at once' (ZBY 07:129, Shi Chong Lao Nu Cai 石崇老奴才). Or insensibility to pain could happen in the condition of unconsciousness and deep sleep: 'I slept soundly for a while and didn't suffer any pain' (ZBY 03:051, Po Yang Hu Hei Yu Jing 鄱陽湖黑魚精).

From the above quick survey of some examples in the linguistic, philosophical, historical and literary fields, we can formulate a tentative general framework of the perception of pain in Chinese culture, which conforms to a universal perception: the close connection of body dynamics with psychophysical evolution and the negative evaluation of such experiences, even if there are attempts to 'rationalise' and explain such phenomena. Evidently, the interconnection between body and heart-mind is more strongly emphasised than in modern Western cultures, maybe because of the lack of a Platonic imprint of soul-body dichotomy. The medical theoretical elaborations are in this direction.

Summing up, we can briefly draft some conclusive observations. Pain experience was different from other feeling sensations because it was associated with a subjective component of emotion and behavioural expression, namely suffering, and, moreover, it might have involved more than one sensation. There were no identical pains, as they acquired peculiar characters and intensity according to: a) circumstances; b) the subject, his/her age and sex, and his/her psychophysical conditions; c) the cultural background. A stimulus that would generally have produced a strong protest of pain could provoke quite a different response in other subjects, with a positive rather than a negative value – and this not only in cases of sadism and masochism, but rather because of the meanings given by individuals to such experiences.

Research in pain pathology shows that bodily and emotional unpleasantness enforce each other. In early China, as in other cultures, physical pain was viewed as a dysfunction of the body caused either by intrusion of a foreign substance or object into the body or by an imbalance of the vital body energies. Similarly, psychological pain was conceived of as a self-generated energetic block that evoked such emotions as anxiety, depression and anger, which are all present in biophysical pain. The experience of pain comprised, most importantly, the complex emotions of anxiety and depression. These concepts partially overlapped with sadness and anger. Anxiety generally is a response to the threat of future harm. It is typically associated with the anticipation of pain (body harm) or loss (separation) and involves immobilisation of coping processes with the core theme of unspecific, vague, objectless existential danger.

The line traditionally drawn between bodily sensation and emotion is maintained with difficulty, as evidenced by a variety of sources, from literature to philosophy. Emblematic of these is the Mencian 'heart of *ceyin* (compassion)', a heart that can be pained by the sight of the suffering of the other – the feeling pain for others' pain. The perception of interaction of physical and moral pain in Chinese culture can be traced back to the basic medical assumption of the unity of emotive and physiological evolution. Thus, affective functions were not separated from the rational abilities of the heart-mind. Descriptive representations in novels and poems, as well as philosophical elaborations, seem to confirm such a notion. As the emotional pain of the heart-mind cannot be separated from bodily pain, terms such as *tong* 痛 can be rendered with the English equivalent 'pain', which may also be applied to descriptions of both so-called bodily conditions and so-called mental conditions.

Abbreviations

GJXS	*Gujin xiaoshuo* 古今小説 (*Yushi mingyan* 喻世明言) (Ancient and Modern Tales)
HLM	*Honglou meng* 紅樓夢 (Dream of the Red Mansion)
JSTY	*Jingshi tongyan* 警世通言 (Common Words for Warning the World)
LK	*Shenyinyu* 呻吟語 (Groaning Words)
LZZY	*Liaozhai zhiyi* 聊齋誌異 (Strange Tales from the Leisure Studio)
MDT	*Mudanting* 牡丹亭 (The Peony Pavilion)
Qingshi	*Qingshi leilüe* 情史類略 (Anatomy of Love)
ZBY	*Zibuyu* 子不語 (What the Master Would Not Discuss)

Notes

1. Few studies have been done on this subject, with the exception of the valuable anthropological essays on contemporary China by Arthur Kleinman, such as *Social Origins of Distress and Disease: Depression and Neurasthenia in Modern China* (New Haven: Yale University Press, 1986), and on the traumas of the Cultural Revolution, 'Pain and Resistance: The Legitimation and Relegitimation of Local Worlds', in Mary-Jo Del Vecchio Good, Paul E. Brodwin, Byron J. Good and Arthur Kleinman (eds), *Pain as Human Experience: An Anthropological Perspective* (Berkeley: University of California Press, 1992), 169–97. See also Eric Hayot, *The Hypothetical Mandarin: Sympathy, Modernity, and Chinese Pain* (Oxford University Press, 2009). Specific topics are discussed in clinical articles (such as Yu Sheng-Yuan, Cao Xiu-Tang, Zhao Gang, Yang Xiao-Su, Qiao Xiang-Yang, Fang Yan-Nan, Feng Jia-Chun, Liu Ruo-Zhuo and Timothy J. Steiner, 'The Burden of Headache in China: Validation of

Diagnostic Questionnaire for a Population-Based Survey', *Journal of Headache and Pain*, 12(2) (2011): 141–6) or in psychological periodicals (such as on the relation of pain and desire for money, Xinyue Zhou, Kathleen D. Vohs and Roy F. Baumeister, 'The Symbolic Power of Money: Reminders of Money Alter Social Distress and Physical Pain', *Psychological Science*, 20(6) (2009): 700–6; and Xinyue Zhou, Cong Feng, Lingnan He and Ding-Guo Gao, 'Toward an Integrated Understanding of Love and Money: Intrinsic and Extrinsic Pain Management Mechanisms', *Psychological Inquiry: An International Journal for the Advancement of Psychological Theory*, 19(3–4) (2008): 208–20).

2. See, for the West, Roselyne Rey, *Histoire de la douleur* (Paris: Editions La Découverte, 1993).

3. Robert Solomon writes: 'A pain suffered without explanation is a different experience from the pain suffered voluntarily, say, in an initiation rite in which one is called upon to demonstrate one's physical courage' (Robert Solomon, 'Some Notes on Emotion', *Philosophy East & West*, 45(2) (1995): 171–202). Moreover Kleinman notices that: 'Chronic pain challenges the simplifying Cartesian dichotomies that still are so influential in biomedicine and also in North American culture ... "real" (that is, physical) versus "functional" (that is, psychological, therefore imaginary) categories' ('Pain and Resistance', 169).

4. See Marcel Mauss, 'Rapports réels et pratiques de la psychologie et de la sociologie', Communication présentée le 10 janvier 1924 à la Société de Psychologie, *Journal de Psychologie Normcle et Pathologique*, 21 (1924). On the side of neuropsychology, an essay by Cytowic stresses the elusive and intricate relationships between emotion, consciousness and self-awareness, the interaction of cognition, emotion and behaviour and their neural substrates, the primacy of emotion and its continual influence on cognitive processes: R.E. Cytowic, *The Man Who Tasted Shapes: A Bizarre Medical Mystery Offers Revolutionary Insights into Emotions, Reasoning, and Consciousness* (New York: Jeremy P. Tarcher/Putnam, 1993). In China, physical pain, associated with illness of the inner organs or the skin, was one of the image domains of metaphors and metonymies to describe psychic suffering, stressing social stigmatisation and marginalisation.

5. See Elaine Scarry, *The Body in Pain: The Making and Unmaking of the World* (New York: Oxford University Press, 1985). An extreme way of communicating suffering is suicide (*ti* 體), which has been studied under various aspects and would require a separate analysis. To take the example of Li Miaohui 李妙惠, the character of the story 'Lu Mengxian searches for his wife on the Changjiang' (盧夢仙江上尋妻) from the *Shi dian tou* by Langxian 浪仙, which was inspired by a real event that happened in the early Ming and that was quite popular among literary circles. Here the end of life was caused by moral suffering, due to the female character's determination to preserve her chastity, but it interacts with the rituality concerning her body and the form of passing away: 'She pondered that there are no more than three ways to die: the knife injures to death the body bequeathed from parents 刀上死傷了父母遺體; by dying in a river, the corpse will float in the waves 河裡死屍骸飄蕩; hanging is much cleaner 不如縊死倒得乾凈.' See *Shi dian tou*, 2:23–52. See also the *wenyan* versions of the tale recorded in *Qingshi* 情史, 'Li Miaohui'. See too *Qingshi*, in *Feng Menglong quanji* (Shanghai: Shanghai

guji, 1993), xxxvii–xxxviii, 13–17. This episode is quoted and commented upon by Barbara Bisetto, 'Perceiving Death: The Representation of Suicide in Ming Vernacular Literature' in Paolo Santangelo (ed.), *From Skin to Heart. Perceptions of Bodily Sensations and Emotions in Traditional Chinese Culture* (Wiesbaden: Harrassowitz, 2006), 155–9.

6. Cf. Janis H. Jenkins, 'Culture, Emotion, and Psychopathology', in Shinobu Kitayama and Hazel Rose Markus (eds), *Emotion and Culture. Empirical Studies of Mutual Influence* (Washington DC: American Psychological Association, 1994), 307–35; A. Kleinman, 'Anthropology and Psychiatry: The Role of Culture in Cross-cultural Research on Illness', *British Journal of Psychiatry*, 151 (1987): 447–54; A. Kleinman, *Rethinking Psychiatry: From Cultural Category to Personal Experience* (New York: Free Press, 1988); A. Kleinman, *The Illness Narratives: Suffering, Healing and the Human Condition* (New York: Basic Books, 1988).

7. Expressions of pain include crying, lamenting (Shen cried out his grievance and yelled for pain. ZBY 09:171, Cheng Huang Shen Xu Jiu 城隍神酗酒), and frowning (for instance, Shen dreamed that the tall man appeared to him, knitting his brows for pain (after punishment with 40 strokes). ZBY 04:076, Chang Gui Bei Fu 長鬼被縛).

8. Angelika Messner, 'Body and Bodily Sensations within an Early Field of Expertise Knowledge in China' in Santangelo (ed.), *From Skin to Heart*, 58.

9. See *Mengzi*, Liang Hui Wang 梁惠王上, 7, which describes the dialogue between the philosopher and a king on the feeling of not bearing to see the suffering of other beings. For the interpretation of the passage, see Wu Xiaoming, '"The Heart that Cannot Bear … the Other": Reading Mengzi on the Goodness of Human Nature', in Santangelo (ed.), *From Skin to Heart*, 165–79. This chapter faces the 'natural' link between sensations and the heart on the basis of the concept of the 'heart that can bear' and the 'heart that cannot bear (the suffering of) others', according to the dialectical relation of 'moving-not moving': as a human being is a corporeal being, from the very beginning of one's life, one already has to bear the hunger, thirst, fatigue, disease, wounds and the ageing of the body, and can understand the suffering of other beings.

10. Michael Nylan, 'On the Politics of Pleasure', *Asia Major*, 14(1) (2001): 73–124 at 94.

11. Cited in P. Santangelo, 'Additional Data Concerning Bodily Sensations and Emotions in Pre-modern Chinese Literature', in Santangelo (ed.), *From Skin to Heart*, 285.

12. For the Yangzhou massacre, see Wang Xiuchu 王秀楚, 'Yangzhou shiri ji' 揚州十日記, in Ming ji baishi chubian 明季稗史初編, comp. Liuyun jushi 留雲居士 (facsimile reproduction of 1936 Shangwu edition; Shanghai: Shanghai shudian, 1988). For these interpretations of Qian's actions, see Yang Jinlong 楊晉龍, 'Qian Qianyi shixue yanjiu' 錢謙益史學研究 (MA thesis, Kaohsiung Normal College, 1989), 5; and Liu Shinan 劉世南, Qingshi liupai shi 清詩流派史 (Taipei: Wenjin chubanshe, 1995), 72–94, quoted by Wai-yee Li, introduction, in Wilt L. Idema, Wai-yee Li, and Ellen Widmer (eds), *Trauma and Transcendence in Early Qing Literature* (Cambridge, MA: Harvard University Asia Center, 2006). See also Angelika C. Messner, 'Towards a History of the Corporeal Dimensions of Emotions: The Case of Pain', *Asiatische Studien/ Études Asiatiques*, 66(4) (2012): 943–72.

13. 'Shen Xiao guan – *niao hai qi ming*' 沈小官 – 鳥害七命 (Shen Xiu Causes Seven Deaths with One Bird), *Gujin xiaoshuo* 古今小說 (Stories Old and New), *juan* 26 (1958), ii, 394, quoted by Messner, 'Towards a History'.

14. In a case of rage, another famous novel, *Jin Ping Mei*, mentions the moving of the five viscera's energy: 'The breaths of his Five Viscera ascended to Heaven' 五臟氣沖天 (Chapter 11).

15. There are many examples of pain in sickness. For some of them, only a physical cause is mentioned: 'Really, on account of her age, the old lady was unable to endure her grief and the vicissitudes of life, and that night she had a headache, a pain in her chest and sore throat, and found difficulty in breathing' (HLM 64:980); 'Lin Daiyu would have gone on writing, but she felt her whole body burning hot, and her face heating' (HLM 34:505). For continuous abdominal pain 腹痛不止, see HLM 69:1051.

16. He Zhao Guangyuan tibi 和趙光遠題壁, Quan Tang shi 全唐詩, poem 802 (1986), 1967.

17. *Chengyu* 成語, from the poem of the Tang 唐 poet Li He 李賀, 'Jin tong xian-ren ci Han ge' (金銅仙人辭漢歌).

18. Old age is also a possible cause of pain: 'Please be indulgent with me, since I am so old my bones are aching quite badly! Thus, please excuse me if I am rather impudent!' 恕我老了骨頭疼, 容我放肆些 (HLM 53:808). 'Really, on account of her age, the old lady was unable to endure her grief and the vicissitudes of life, and that night she had a headache, a pain in her chest, and difficulty in breathing' 果然年邁的人, 禁不住風霜傷感, 至夜間便覺頭悶心酸, 鼻塞聲重 (HLM 64:980).

19. From the Chinese medical point of view, concerning the more general melancholy, it was interpreted exclusively as a disease of the body and of the mind. See, for example, the pages of the *Handbook of Pharmacology* (*Bencao gangmu* 本草綱目, 12:93–4) by Li Shizhen 李時珍 (1518–93), referring to certain properties of ginseng. On Chen Shiduo's 陳士鐸 (1687) thesis concerning a kind of feminine hysteria owing to lovesickness and excessive passion, *huadian* 花癲, see Angelika Messner, *Emotions in Late Imperial Chinese Medical Discourse: A Preliminary Report* (Naples: Ming Qing yanjiu, 2000).

20. See LZZY 2:147.

21. See *Qingpingshan tang huaben* 20:249. Lovesickness was actually considered the cause of a series of physical disorders, such as insomnia and loss of appetite, as we can see in several examples: 'the young scholar thought ardently of the woman, and consequently forgot completely to eat and sleep' 生念女蒙切, 寢食俱廢 (*Xiangyinlou bintan*, 1:3); 'love-death' or 'suicide for love', *qingsi* 情死, are both the fatal end of an unhappy love. Linked to this concept is that of *xiangsisi* 相思死 'death by love-sickness', whose reappraisal in the late Ming and Qing literature emerges from several phrases like 'only those who in the world are passionately in love can fully understand what death by love means' (*tianxia you qingren, jin jie xiangsisi* 天下有心人盡解相思死). This proposition is based on that of the *Taiping guangji* (160:37 [3:321], *Hou Jitu* 侯繼圖) 'He who in this world is ungrateful and heartless understands fully what death by love means' (*tianxia fuxin ren, jin jie xiangsisi* 天下負心人盡解相思死), and is found in numerous Ming and Qing works, such as the 'Anatomy of Love' (*Qingshi, Hou Jitu*, 2:48) by Feng Menglong, or the 'Biography of Fan Muzhi' (*Fan Muzhi zhuan, Zhenzhu chuan* 珍珠船 'The

pearl ship', *juan* 1) by Chen Jiru. Cf. also Gōyama Kiwamu合山究, 'Min Shin jidai ni okeru jōshi to sono bungaku' 明清時代における情死とその文學, *Itō Sōhei kyōju kinen Chūgokugaku ronshū* (1986), 417–49, especially 448n4. In one famous verse, Tang Xianzu sang of the suffering due to love, which can lead to death by suicide: 'Why die because of amorous passion (*siqing* 死情)? There must be something divine in suffering' (*Tang Xianzu ji*, 1:655).

22. The text of the *Honglou meng* that is utilised here is the electronic text kindly provided by the publisher Zhonghua shuju, thanks to the help of Professor Guo Yingde, who cooperates with my research project in the textual analysis of the work. The original edition is *Chengjia* 程甲本 (printed in 1791, e-reprinted in Beijing by the Publisher Zhonghua shuju, 1998).

23. 'I remember when my illness started. I was standing up feeling quite all right, but then it seemed as if someone had hit my head from behind with a stick, and the pain was so bad that in my eyes everything went black. Still, I could see evil ghosts with dark faces and protruding teeth, all over the place, who were swinging swords and clubs. When I lay down on the *kang*, I felt as if I had tight bands around my head, and the pain became so acute that I lost consciousness' (HLM 81:1238).

4
Psychological Pain: Metaphor or Reality?

David Biro

In the opening chapters of *Anna Karenina*, Dolly discovers that her husband Stiva has cheated on her again. She is furious yet also heartbroken. Tolstoy writes that Dolly 'winces as if from physical pain'. He repeats this several times for emphasis: 'Dolly again feels pain and wishes she could inflict even a tiny bit of the same physical pain on him'; 'She cried out, not looking at him, as if the cry had been caused by physical pain.'[1]

While Tolstoy's characterisation here illustrates his profound insight into human nature, it also points to a confusion about pain that persists today. The fact that Tolstoy uses the simile form and finds it necessary to qualify the pain as 'physical' suggests that he may not be so sure about what his character is experiencing. Is it really the *same* feeling one has after breaking a leg or suffering a burn even though nothing has happened to Dolly's body? Or merely *like* that feeling in some respect or another?

The question raised by Tolstoy's novel is the subject of this chapter. Can betrayal or rejection lead to pain in the sense that we understand physical pain? How about the death of a loved one? Or a person in the grip of a severe depression? Ultimately, I ask whether these affective states – construed as injuries to the mind – can trigger the same kind of pain as injuries to the body.

Background

Historically, scientists and physicians would have answered this question with a resolute no. Since the time of Descartes, pain (in the West) has been understood as a strictly physical phenomenon.[2] It occurs when receptors on nerve cells in the skin and internal organs detect damaging stimuli to the body, a pin-prick, for example, or high temperatures. The nociceptors (from the Latin *nocere*, to injure) signal the brain, which responds, in turn, with a series of protective measures. We pull the arm away from the flame and rest the broken leg. This highly effective biological warning system is critical to survival.

The absolute connection between pain and physical injury, however, was called into question during the second half of the twentieth century. Researchers observed occasions when there was devastating injury (wounded soldiers on the battlefield) and yet little or no pain and, conversely, occasions when minor injury produced excruciating pain (migraine).[3] In addition, a variety of psychological factors – emotions, expectations, attitudes and memories – were shown to be capable of significantly modifying pain experience.[4] These findings would eventually be explained by the paradigm-shifting gate control theory of pain introduced by Melzack and Wall in 1965.[5] Thereafter, the simple stimulus-response model of pain was replaced by a more complex perceptual system whereby nociceptive signals could be influenced at multiple points along their pathways to and from the brain.

In order to accommodate this more nuanced understanding of pain, a new definition was needed. Psychiatrist Harold Merskey proposed the following: 'Pain is an unpleasant sensory and emotional experience associated with actual or potential tissue damage or described in terms of such damage.'[6] The formulation was subsequently taken up by the International Association for the Study of Pain (IASP) and is the most widely circulated definition today. But while Merskey adds an emotional or affective component to pain and dilutes its connection to physical injury by adding the notion of 'potential damage', he still does not satisfactorily address the problem of psychological pain. In the case of betrayal, grief or depression, there is no tissue damage (either actual or potential), nor is it a matter of an emotional state influencing how tissue damage is experienced. The point is that pain in these instances is caused *solely* by psychological damage. But since the IASP definition and the prevailing scientific models do not allow for such a sequence of events – pain remains bound to physical injury and nociceptor pathways – psychological 'pain' remains outside the scope of pain proper. In fact, most contemporary scientists and physicians would argue that it belongs to a different category altogether, which should more appropriately be labelled suffering or anguish.[7] Hence, there is no mention of grief or depression in medical classification schemes of pain.

Even psychiatrists are wary of speaking about psychological pain in their domain. Pain of the physical variety can accompany, exacerbate and in some instances cause psychiatric illness.[8] But the reverse – a psychiatric illness, for example, directly causing physical-like pain that is unrelated to physical injury – is not commonly accepted, except perhaps in the relatively rare case of psychogenic pain, the modern-day equivalent of what Freud once termed hysteria or conversion reaction.[9] The bottom line is that psychological pain is an oxymoron and is at best a metaphor.

The problem with this view is that it does not square with the way most laypeople feel and express their feelings, and have been doing so at least since Tolstoy's day. Psychological or emotional injury that occurs in the setting of betrayal, grief and depression are routinely described as painful. Moreover, such subjective feelings and their expression are the gold standard by which pain is evaluated and measured. Despite advances in neurobiology, there is no definitive, objective way to assess pain. Even if a relatively sensitive and specific neurological signature of pain in the brain has been discovered, as a recent *New England Journal of Medicine* report suggests,[10] it will never be the equivalent of (or replace) what is essentially a subjective experience.[11]

So when Dolly and others say they feel pain, how can we ignore them? Either we agree with the scientist/physician that they are mistaken and what they are feeling is not pain but something categorically distinct, or we must acknowledge pain's presence and change, once again, what we mean by the word and concept of pain.

Like Tolstoy, I believe that pain can indeed occur outside the setting of physical injury and that we must therefore further broaden our definition to accommodate these instances. Two threads of evidence are presented in support of this view: linguistic and neuroscientific.

Linguistic evidence

As mentioned earlier, most people use the word 'pain' when they break a leg and when they lose a spouse – that is, they use the same label for the feeling that accompanies both physical and psychological injury. In addition, they talk of hurt, ache and suffering in both cases.[12] This practice is not exclusive to English speakers and can be observed across a wide variety of languages and cultures, from Hungarian to Inuktitut.[13] There is also a tendency amongst sufferers to compare these qualitatively interchangeable states more quantitatively, with psychological pain typically winning out in terms of intensity. Of the 30 depressed patients in one study, all of whom also had a history of a life-threatening physical illness, 28 considered their psychological pain worse than any physical pain they experienced.[14]

More importantly, when asked to be more descriptive, to try and communicate how such experiences feel, psychological and physical pain sufferers will do so in similar ways. Pain of any kind is notoriously difficult to express. There are problems conceptualising the experience because it is perceptually inaccessible (we cannot see or touch pain) and because, unlike other subjective states, it is not always linked to external objects that we can see (for example, the person who makes us angry or the pit-bull that frightens us).[15] As a result, we are forced to think of pain indirectly, through metaphor: we imagine a more knowable object connected to the pain and then speak of the experience in terms of that object. We speak of the private and invisible in terms of the public and more accessible.[16]

By far the most common metaphor used to describe pain is the weapon.[17] We say a pain is shooting or stabbing. Lengthy lists of similar adjectives can be found on the McGill Pain Questionnaire, which was created in the 1970s by Ronald Melzack and Gil Torgerson to help patients communicate their pain to doctors: piercing, drilling, burning, grinding, throbbing, stinging, squeezing and so on. Each of the descriptors implies the presence of a weapon or weapon-like object that can injure the body – the drill that *drills*, the fire that *burns*. And since most patients have never been stabbed or shot, they are using these terms figuratively to objectify their experiences; now they can *see* pain and describe how they feel by talking about knives and guns, and the damage they can do to the body.

It turns out that people with psychological pain use the very same metaphors to describe their experiences.[18] Wracked with grief by the loss of her husband, Joan Didion envisions giant waves. In her memoir, *The Year of Magical Thinking*, she writes that she felt as if she were being battered by 'destructive waves, paroxysms, sudden apprehensions that weaken the knees and blind the eyes and obliterate the dailiness of life'.[19] Waves in their temporal and physical dimensions are weapon-like objects that move toward and strike the body. In his classic paper on grief, psychiatrist Eric Lindemann found that the majority of bereaved subjects he studied routinely experienced such destructive waves.[20]

The Mexican painter Frida Kahlo often felt like Tolstoy's Dolly. In *Memory* (1937), she depicts her pain from the repeated infidelities of her husband Diego Rivera as a sword that pierces her vest and penetrates the left side of her chest. In the painting's foreground lies the proverbial broken heart, greatly enlarged, detached from its normal place in the body and spurting blood from every ventricle. Tears stream down Frida's cheeks while a threatening sky looms overhead.

Novelist William Styron suffered from depression and wrote about it in his memoir *Darkness Visible*. When the pain was at its worst, Styron felt like he was being 'suffocated' and 'drowned', like a 'howling tempest was battering his brain'.[21] Kay Redfield Jamison, a psychiatrist also battling depression, imagines a more elaborate weapon for her pain: a giant centrifuge, containing tubes of blood, that spins around her mind faster and faster, until it explodes and splatters blood everywhere.[22]

Listening to the language of pain of all varieties – in the clinic as well as in the arts and literature – we discover a shared felt structure that the weapon metaphor effectively captures. Whether triggered by grief and depression or kidney stones and spinal injury, pain reads like a narrative in three parts:

Weapon → Injury → Withdrawal

In pain we feel as if there must be some weapon-like object (wave, sword, centrifuge) that moves toward and threatens us; that when it strikes, it will cause injury; and from which we must turn away. Even when there is nothing moving against us, when there is no injury, when we remain stationary, we *feel* the movement, the injury and the desire to run.[23] And because those same feelings are present in both psychological and physical injury, people naturally label the experience with the same word and describe it using the same metaphors.[24]

Evidence in the brain

There is also new evidence for broadening our notion of pain to include instances of psychological injury. As previously mentioned, the introduction of gate control theory progressively weakened the link between tissue damage and pain so that we can no longer understand pain in terms of the body alone. We are now very much aware that a host of extracorporeal factors – one's culture and past experiences, our emotional and cognitive states, the context of pain – can intensify or dampen a nociceptor signal before and after it registers in higher brain centres.[25] Moreover, many cases of chronic pain seem to occur without any direct nociceptor stimulation at all. Neuropathic pain, for example, results when a dysfunctional nervous system fires spontaneously or misinterprets harmless sensory stimuli as noxious.[26] In some cases, neuropathic pain follows on the heels of a specific physical injury, while in others, no preceding injury can be identified. In trigeminal neuralgia or *tic douloureux*, the movement of a feather across the face can trigger spasms of intense pain.[27]

A second strand of evidence comes from our growing understanding of how the brain processes pain. We have learned that pain is a highly complex perceptual system with multiple subsystems. Most important for this discussion are the distinct areas in the brain that process the sensation of pain (its quality, location and intensity) and our feelings about the sensation (the narrative of its aversiveness).[28] Typically, the *sensory* centre (in the somatosensory cortex) and the *affective* centre (in the anterior cingulate and insula cortices) are linked and activated simultaneously: tea spills on the arm, generates a burning sensation that is felt to be damaging and initiates a series of protective responses.

However, in certain instances, these centres can be unlinked. For example, a person can have the sensation of pain but not feel pain.[29] This is observed in patients undergoing minor surgery with medication that makes them indifferent to an incision made with a scalpel. Even more dramatic is a rare group of patients with *pain asymbolia*, whose affective pain centres (or the connections to those centres) have been destroyed. These patients can sense a needle prick (because the nociceptor signal registers in the somatosensory cortex) but will laugh at its insignificance (because the signal is not processed by the anterior cingulate cortex).[30] Such cases hardly resemble what we think of as pain because without the *feeling* of pain, protective measures will not be taken and the experience loses its biological significance. In other words, without the affective component, pain becomes meaningless.

What about the reverse, namely having the feeling of pain without specific sensations of pain generated via nociceptor pathways? There is now evidence that affective pain centres in the brain can be directly activated by psychological injury. Naomi Eisenberger and her colleagues at UCLA have recently developed a clever model of what they call social pain, namely the painful feelings that follow social rejection or loss.[31] Subjects were asked to play a video ball-tossing game while their brains were monitored by fMRI. At a certain point, the subjects were excluded from the virtual game and reported experiencing distress that correlated with increased blood flow to the anterior cingular and insular cortices.[32] This is exactly the same pattern that would have occurred had they been pricked by a needle (except for the absence of somatosensory cortical blood flow, which was expected since there was no tissue damage). The greater the social pain generated, the more active the affective pain centres became. Similar studies were carried out on grieving subjects and they showed the same results.[33] In grief as well as rejection, people *feel pain* which is reflected in their brain scans and in the words they use. Thus, the most meaningful component of pain appears to be fully operative in the absence of physical injury.[34]

Pain as feeling

Clearly the most critical aspect of pain from a biological point of view is its affective component, the aversive feeling of injury or impending injury. In fact I would argue that when tissue damage is present but not felt as pain (wounded soldiers, anaesthesia, *pain asymbolia*), we should not label the experience as 'pain' at all. On the other hand, when pain is felt in the absence of tissue damage (in the cases of Tolstoy's Dolly, Frida Kahlo, Joan Didion and Kay Redfield Jamison given above), the pain is very real indeed and serves the same biological signal as physical pain, motivating sufferers to take protective measures.

Pain is fundamentally an alarm system that has evolved to protect us from injury. In earlier times the threats were primarily physical. Primitive pain pathways are found in single-celled creatures like the paramecium so that it could avoid noxious physical stimuli.[35] At some point in evolution, however, when consciousness and self-awareness developed, the nature of potential threats would have naturally expanded to include noxious psychological stimuli, and physical pain pathways might have been used to regulate those threats. This is precisely what seems to have happened in the case of separation distress. In his studies on a wide variety of non-human mammals (dogs, guinea pigs, chicks, rats and primates) separated from their mothers during infancy, Jaak Panskepp found that physical pain mediators like morphine and other opioids were able to alleviate this patently non-physical pain (as measured by the isolation cries of the animals).[36]

Clearly social bonds had become critical to mammalian well-being and survival, requiring mechanisms to recognise and react to the threat of exclusion.[37] The same is likely true for all aspects of a human being's psychological integrity. Because our conscious, inner lives are now as important to our well-being as our bodies, we must have ways to protect ourselves from psychological injury. Researchers including Panskepp and Eisenberger have proposed that over the course of evolution, psychological pain has 'piggybacked' onto the pre-existing physical pain alarm system, borrowing its signals and mediators to preserve our psychological health.[38]

I would also argue that this borrowing between the two pain systems (as well as the progressive integration of the mental and physical spheres in general) has led to a progressive blending and blurring of

pain experience, whereby it has become increasingly difficult to determine what kind(s) of injury produces our pain. We are very far from the
single-celled creature that responds to threatening stimuli in a reflexive,
unfeeling way. In fact, at this point there may be no such thing as an
isolated physical pain, just as there may be no such thing as an isolated psychological pain. Pain is always a composite. Cancer patients
naturally have pain from primary and metastatic tumours, but they also
experience the psychological pain of overwhelming fear and threat.[39]
Likewise, Joan Didion and other grieving subjects typically complain
of physical symptoms that can include difficulty breathing, fatigue and
tightness in the throat.[40]

An interesting study on social pain illustrates this blending of the
physical and psychological and its neural correlates. Researchers asked
subjects to relive an experience of rejection while they were shown
pictures of a boyfriend or girlfriend who recently broke up with them.
Their intention was to generate a more intense pain than Eisenberger's
Cyberball model. As expected, fMRI scanning revealed activation of the
affective pain centres in the brain. However, there was also activation
of somatosensory centres (even though nociceptors were presumably
silent).[41] These findings begin to show how psychological injury might
lead to the physical symptoms reported by so many psychological
pain sufferers. The bottom line is that pain – its causes and its felt
manifestations – will always involve both the body and the mind.[42]

Consequences of a broader approach

The time is ripe for broadening our definition of pain, recognising it as
the feeling of injury to a *person* rather than a body. Such a reformula-
tion will gradually lead to changes in the way we understand and manage pain. First, it will reduce the semantic confusion involved in pain
language so that we will not constantly need to qualify and pigeonhole
pain (as Tolstoy did in *Anna Karenina* and as I have done throughout
this chapter) as either physical or psychological (or even social), but
regard it instead as a composite. Whenever the aversive feeling of injury
or the threat of impending injury is experienced by a person, there will
inevitably be pain.

Second, a broader understanding of pain would equalise the different
types of pain, which is not only intellectually important but morally
so. The traditional, dualistic paradigm privileges physical pain: if there
is no tissue damage or lesion on fMRI, then there is no 'real' pain. But
how then should we respond to Frida Kahlo, Joan Didion and countless

less well-known sufferers who insist that the pain they feel is real and in many cases is a lot worse than any physical pain they have experienced. Indeed, suicide rates are significantly higher in the setting of grief and depression than they are in the setting of physical pain.[43]

In addition to relegating psychological pain to second-class status, the traditional paradigm is also harmful to another large (and growing) group of sufferers.[44] Patients with chronic pain from migraine, lower back conditions and fibromyalgia find themselves in limbo between 'real' pain and the derivative variety. On the one hand, their pain seems physical (because it is localisable to the body), but, on the other hand, it has more in common with the psychological kind (because there is no detectable tissue damage). Not surprisingly, medicine has been ineffective at managing such patients. Worse, their pain is often not believed. Although things have improved for chronic pain patients with the advent of pain specialists and pain clinics, many are still tormented by the insidious logic of the prevailing biomedical approach.[45]

Finally, a change in mindset would encourage new approaches in treating pain. One might, for example, offer 'physical pain' medication for 'social pain', as DeWall and associates did when they administered acetaminophen (paracetamol) to subjects and found that it reduced their daily complaints of distress (along with anterior cingulate cortical activity in response to exclusion from the video ball-tossing game).[46] By the same token, 'psychological pain' therapies might be administered to treat cancer and chronic pain since we know from placebo studies that belief and expectation are effective analgesics.[47] Similar levels of pain relief have occurred by bolstering a patient's social support and inducing pleasurable feelings. A recent study, for example, found that showing a subject a picture of a romantic partner was able significantly to reduce the intensity of a painful stimulus.[48]

Towards a new definition

Ultimately, we must move away from the old definition of pain (with its emphasis on tissue damage) and find a new and more useful one. As a starting point, I propose the following: 'Pain is the aversive feeling of injury to one's person and the threat of further, potentially more serious injury. It can only be described metaphorically.' This definition includes five critical elements:

1. Pain is a *feeling*, in neurologist Antonio Damasio's sense of the term. Damasio defines feeling as a higher-order, conscious appraisal

of an organism's state at a given time, an appraisal that prompts self-regulating behaviour aimed at ensuring well-being.[49]

2. The feeling of pain signals an *injury* occurring at the level of the *person*, be it bodily or psychological damage or, more commonly, a composite of the two.

3. The feeling of pain prompts self-regulating behaviour. Pain's inherent *aversiveness* (from the Latin *avertere*, to turn away from) urges us to withdraw the arm from the flame and whatever else we can do to alleviate it.[50]

4. The feeling of pain involves the present as well as the future. Pain signals the presence of injury as well as the *threat* of further, potentially more serious injury.[51]

5. Because of the difficulties involved in conceptualising and representing pain, the feeling of pain can only be described *metaphorically*.

Notes

1. L. Tolstoy, *Anna Karenina* (trans. Richard Pevear and Larissa Volokhonsky) (London: Penguin, 2000), 3, 11.

2. In Vernon Mountcastle's popular textbook, for example, the definition of pain as 'a sensory experience evoked by stimuli that injure or threaten to destroy tissue' (*Medical Physiology* (St Louis: C.V. Mosby, 1980), 391) is essentially a restatement of Descartes' simple stimulus-response model (a flame activates skin particles ... pulls on a cord ... sets off a bell in the brain ... produces pain).

3. R. Melzack and P.D. Wall, *The Challenge of Pain* (London: Penguin, 1996), 3–14.

4. *Ibid.*, 15–33.

5. R. Melzack and P.D. Wall, 'Pain Mechanisms: A New Theory', *Science*, 150 (1965): 971–9.

6. H. Merskey, 'Psychological Aspects of Pain', *Postgraduate Medical Journal*, 44 (1968): 297–306; IASP Subcommittee on Taxonomy, 'Pain Terms: A List with Definitions and Notes on Usage', *Pain*, 3 (1979): 249–52.

7. E. Cassell, *The Nature of Suffering and the Goals of Medicine* (New York: Oxford University Press, 1991), 30–46.

8. M.J. Bair, R.L. Robinson, W. Katon and K. Kroenke, 'Depression and Pain Comorbidity: A Literature Review', *Archives of Internal Medicine*, 163 (2003): 2433–45.

9. The category of psychogenic pain in DSM-III (1987) has since been replaced by the broader category of 'pain disorder' in DSM-IV (2000) and DSM-V (2013), but these changes still fail to address the pervasiveness of pain experienced and reported by psychiatric patients. See Merskey, 'Psychological Aspects of Pain'.

10. T.D. Wager, L.Y. Atlas, M. Lindquist *et al.*, 'An fMRI-Based Neurologic Signature of Physical Pain', *New England Journal of Medicine*, 368 (2013): 1388–97.

11. The failure of the scientific or materialist programme to account for subjective experience like pain has been the concern of philosopher Thomas Nagel since his seminal article 'What is it Like to Be a Bat' *Philosophical Review*, 83 (1974): 435–50. According to Nagel, any psychophysical reduction, like an fMRI signature of pain in the brain, will inevitably leave out the essential character of a subjective experience, namely that it is experienced (and accessible) from only one point of view.

12. S. Mee, B.G. Bunney, C. Reist *et al.*, 'Psychological Pain: A Review of Evidence', *Journal of Psychiatric Research*, 40 (2006): 680–90.

13. G. MacDonald and M.R. Leary, 'Why Does Social Exclusion Hurt? The Relationship between Social and Physical Pain', *Psychological Bulletin*, 131(2) (2005): 202–23. This includes Tolstoy's Russian where the word *bol'* can refer to both physical and psychological pain.

14. H. Osmand, R. Mullaly and C. Bisbee, 'The Pain of Depression Compared with Physical Pain', *Practitioner*, 228 (1984): 849–53.

15. D.E. Biro, *The Language of Pain: Finding Words, Compassion, and Relief* (New York: Norton, 2000), 36–47.

16. *Ibid.*, 79–96.

17. E. Scarry, *The Body in Pain: The Making and Unmaking of the World* (New York: Oxford University Press, 1985), 15–19.

18. Merskey deserves credit for recognising the metaphorical language of pain and, in particular, its dependence on physical injury, even in the case of psychological pain where sufferers invariably talk 'in terms of tissue damage'.

19. J. Didion, *The Year of Magical Thinking* (New York: Knopf, 2005), 27–8.

20. E. Lindemann, 'Symptomatology and Management of Acute Grief', *American Journal of Psychiatry*, 101 (1944): 141–8.

21. W. Styron, *Darkness Visible* (New York: Vintage, 1992), 17, 38.

22. K.R. Jamison, *An Unquiet Mind: A Memoir of Moods and Madness* (New York: Vintage, 1996), 79–80.

23. Biro, *Language of Pain*, 91–3.

24. While the basic structure of the weapon metaphor has remained relatively constant, it has been adapted and elaborated to fit the needs of people in different times and places. See Joanna Bourke, 'Pain and the Politics of Sympathy, Historical Reflections, 1760s to 1960s' (University of Utrecht, 2011), available at: http://dspace.library.uu.nl/bitstream/handle/1874/210217/Bourke_Joanna_oratie.pdf?sequence=1 (date accessed 21 February 2014); and *The Story of Pain: From Prayer to Pain Killers* (Oxford University Press, forthcoming, 2014).

25. Melzack and Wall, *Challenge of Pain*, 15–33; T. Hampton, 'A World of Pain: Scientists Explore Factors Controlling Pain Perception', *Journal of the American Medical Association*, 296(20) (2006): 2425–8.

26. C.J. Woolf and R.J. Mannion, 'Neuropathic Pain: Aetiology, Symptoms, Mechanisms, and Management', *The Lancet*, 353 (1998): 1959–64.

27. J.M. Zakrzewska, *Insights: Facts and Stories Behind Trigeminal Neuralgia* (Gainesville, FL: Trigeminal Neuralgia Association, 2006).

28. D.D. Price, 'Psychological and Neural Mechanisms of the Affective Dimension of Pain', *Science*, 288 (2000): 1969–72.

29. N. Grahek, *Feeling Pain and Being in Pain* (Cambridge, MA: MIT Press, 2007), 29–50.

30. M. Berthier, S. Statkstein and R. Leiguarda, 'Pain Asymbolia: A Sensory-Limbic Disconnection Syndrome', *Annals of Neurology*, 24 (1988): 41–9.

31. N.I. Eisenberger and M.D. Lieberman, 'Why Rejection Hurts: The Neurocognitive Overlap between Social Pain and Physical Pain', in K.D. Williams, J.P. Forgas and W. Von Hippel (eds), *The Social Outcast: Ostracism, Social Exclusion, Rejection, and Bullying* (New York: Cambridge University Press, 2005), 109–27.

32. N.I. Eisenberger, M.D. Lieberman and K.D. Williams, 'Does Rejection Hurt? An fMRI Study of Social Exclusion', *Science*, 302 (2003): 209–92.

33. H. Gundel, M.F. O'Connor, L. Littrell *et al.*, 'Functional Neuroanatomy of Grief: An fMRI Study', *American Journal of Psychiatry*, 160 (2003): 1946–53; M.F. O'Conner, D.K. Wellisch, A. Stanton *et al.*, 'Craving Love? Enduring Grief Activates Brain's Reward Centers', *NeuroImage*, 42 (2008): 969–72.

34. Additional evidence of overlapping neural pathways for social and physical pain include: 1) studies showing that individuals who are more sensitive to one type of pain are also more sensitive to the other; and 2) studies showing that factors that can modulate one type of pain can also modulate the other. See Naomi Eisenberger, 'The Neural Bases of Social Pain: Evidence for Shared Representations with Physical Pain', *Psychosomatic Medicine*, 74 (2012): 126–35.

35. W.R. Clark and M. Grunstein, *Are We Hardwired?: The Role of Genes in Human Behavior* (New York: Oxford University Press, 2000), 34–7.

36. J. Panskepp, B.H. Herman, R. Conner *et al.*, 'The Biology of Social Attachments: Opiates Alleviate Separation Distress', *Biological Psychiatry*, 13 (1978): 607–18.

37. Reviewing data from a number of animal studies, Baumeister and Leary argue that social animals that form strong relationships and are integrated into group living are most likely to survive, reproduce and raise offspring to reproductive age. R.F. Baumeister and M.R. Leary, 'The Need to Belong: Desire for Interpersonal Attachments as a Fundamental Human Motivation', *Psychological Bulletin*, 117 (1995): 497–529.

38. N.I. Eisenberger, 'The Pain of Social Disconnection: Examining the Shared Neural Underpinnings of Physical and Social Pain', *Nature Reviews Neuroscience*, 13(6) (2012): 421–34.

39. N.I. Cherney, N. Coyle and K.P. Foley, 'Suffering in the Advanced Cancer Patient: A Definition and Taxonomy', *Journal of Palliative Medicine*, 10 (1994): 51–70.

40. Lindemann, 'Symptomatology and Management of Acute Grief'.

41. E. Kross, M. Berman, W. Mischel *et al.*, 'Social Rejection Shares Somatosensory Representations with Physical Pain', *Proceedings of National Academy of Sciences USA*, 108(15) (2011): 6270–75.

42. It should also be said – and indeed will be emphasised throughout this book – that pain also always involves the social context and culture in which a sufferer is situated, especially when that sufferer is engaged in representing and finding meaning for his or her pain. Key works that explore the social dimensions of pain include Arthur Kleinman, *The Illness Narratives: Suffering, Healing, and the Human Condition* (New York: Basic Books, 1988); David Morris, *The Culture of Pain* (Berkeley: University of California Press, 1991) and *Illness and Culture in the Postmodern Age* (Berkeley: University of

California Press, 2000); Javier Moscoso, *Pain: A Cultural History* (Basingstoke: Palgrave Macmillan, 2012); and Bourke, *Story of Pain*.

43. E.S. Schneidman, 'Perspectives on Suicidology: Further Reflections on Suicide and Psychache', *Suicide and Life-Threatening Behavior*, 28 (1998): 245–50.

44. The prevalence of chronic pain in the general population is estimated at between 10 and 55 per cent. C. Harstall and M. Ospina, 'How Prevalent is Chronic Pain?', *Pain Clinical Updates International Association for the Study of Pain*, 11 (2003): 1–4.

45. See Lous Heshusius' poignant memoir of her life with chronic pain: *Inside Chronic Pain: An Intimate and Critical Account* (New York: Cornell University Press, 2009); and Deborah Padfield's photographic collaborations with chronic pain patients in *Perceptions of Pain* (Stockport: Dewi Lewis, 2003).

46. N.C. DeWall, G. MacDonald, G. Webster *et al.*, 'Acetaminophen Reduces Social Pain: Behavioral and Neural Evidence', *Psychological Science*, 21(7) (2010): 931–7.

47. F. Benedetti, *Placebo Effects: Understanding the Mechanisms in Health and Disease* (Oxford University Press, 2009), 38–52. See also Moscoso, Chapter 2, this volume.

48. J. Younger, A. Aron, S. Park *et al.*, 'Viewing Pictures of a Romantic Partner Reduces Experimental Pain: Involvement of Neural Reward Systems', *Plos One*, 5(10) (2010): e13309.

49. A. Damasio, *Looking for Spinoza: Joy, Sorrow, and the Feeling Brain* (New York: Harcourt, 2003), 83–133.

50. On first blush, the pain-seeking behaviour of the sadomasochist or religious ascetic might seem to contradict the fundamental aversiveness of pain. However, the goal in these cases is not pain but pleasure, and the only way to achieve that goal – for those who embrace the particular narrative – is *through* pain and all its aversiveness. In a related sense, for many who engage in self-harming behaviour, the motivation is to replace a greater psychological pain with a lesser (and more visible/believable) physical pain.

51. Biro, *Language of Pain*, 99–110. Thomas Szasz, drawing on Freud, was one of the first to think of pain in terms of a threat to the continuity or integrity of the body (for which I would substitute person). See Thomas Szasz, *Pain and Pleasure: A Study of Bodily Feelings* (Syracuse University Press, 1988), 59–62.

5
Phantom Suffering: Amputees, Stump Pain and Phantom Sensations in Modern Britain

Joanna Bourke[1]

The suffering inflicted by the First World War did not end in 1918. When Lieutenant Francis ('Frank') Hopkinson died on 17 December 1974, he was 85 years of age and had lived over half a century in severe pain as a result of having been wounded during the Third Battle of Ypres on 12 August 1917. He had undergone numerous operations, including having his left leg amputated three times. He had also been hospitalised with shell shock. From those terrifying months in 1917 and 1918 until his death in 1974, he endured profound physical and mental anguish due to an agonisingly tender stump and phantom limb pain.

The life of Frank Hopkinson serves as a reminder that the effects of wartime wounding lasted entire lifetimes. After the First World War, millions of men returned home with distressing physical and psychological wounds. Their lives were ruled by pain, despair and conflict with the authorities and medical personnel. Although their continued suffering was often dismissed or treated as inauthentic, disabled service personnel could not simply shrug off their misfortune; young lives could not simply be resumed. The war-afflicted body in pain was a life sentence.

Hopkinson's life can also be used as a lens through which to reflect on two debates within British society. The first relates to the relationship between lesions and suffering. Hopkinson experienced severe physical pain in a limb that had no physiological existence (phantom limb pain) and because his physicians believed his account of pain, they sought a 'cause' in some kind of pathology of his stump. The second debate arose when his doctors failed to discover any underlying biological pathology; this led them to posit an emotional basis for his suffering. In part, this latter shift was due to frustration among the physicians about being unable to ease his pain. However, it was also prompted by a broader trend towards a psycho-social model of pain from the late 1930s onwards.

I will be arguing that a close study of the painful experiences of Frank Hopkinson – whose life as a limbless ex-serviceman spanned most of the twentieth century (1917 to 1974) – can shed a light on the responses of the medical profession in Britain to the physiology and psychology of acute and chronic pain more generally. In other words, this chapter is an exercise in microhistory, or the close study of one individual in order to reflect on broader responses within British society. As historian Filippo de Vivo has observed, microhistorical approaches to history act as 'an antidote to the teleology and elitism of traditional political history' and serve as 'an alternative to the reductive determinism' of some forms of social history.[2]

The specificity as well as the heterogeneity of the lives of so-called 'ordinary' people turns out to be an extraordinary frame through which historians can reflect on general trends. In other words, an examination of one man's life serves as a lens through which the broader culture – including local, national and even global contexts – can be illuminated. After all, Hopkinson's life was profoundly affected by his immediate surroundings: damp weather (which would make his phantom limb pulsate) and hearing hymns (which made him cry) were significant events in his world. So too were national events. Obviously, the British government's declaration of war disrupted everything in Hopkinson's world, but, subsequently, national debates about pensions, innovations in artificial limb technologies and global economic trends proved decisive. As historian Seth Koven has observed, disabled First World War veterans were 'dismembered persons in a literal sense but also in a social, economic, political, and sexual sense'.[3] The physicians who treated Hopkinson operated within a medical culture influenced by specifically British factors (the establishment of the Ministry of Pensions in 1916 and, later, the National Health Service, for instance), but they were also embedded within global scientific communities. An exploration of Hopkinson's life sheds light on the treatment of men wounded during the First World War, as well as on scientific and medical understandings about the nervous system in general, and on stump and phantom pain in particular.

Dismemberment

Who was Frank Hopkinson? He was born in 1889 into a privileged family. He was the second son of Canon Charles Girdlestone Hopkinson,

Rector of Whitburn (in Sunderland), and was educated at Marlborough, an independent boarding college dedicated to schooling the sons of Church of England clergy. This strapping, six-foot-tall young man who enjoyed riding horses was reputed to be 'not very bright',[4] so, after a spell working as a clerk in a Nitrates Works in northern Chile, he eagerly returned to join the 11th Durham Battalion a few days after the declaration of war. At that time, medical officers judged him to be 'not particularly nervy', with 'normal' health.[5]

This all changed on 12 August 1917 when a bomb dropped from a plane smashed his left leg into fragments. His leg became infected with gas gangrene and had to be amputated – once in the No. 24 General Hospital at Etaples and then a second time later that year at the King Edward VII Hospital for Officers. He underwent a third re-amputation in 1927.[6] Hopkinson was distraught: 'It has been found necessary to reamputate the femur for protruding bone, as there is no cushion or protection whatever to the end of the stump.' This was a serious setback since 'I am now faced with the prospect of another operation, leaving very little stump, at a time when I had hoped to start work'.[7] He never experienced prolonged periods of employment again.

The site of his amputation was particularly unfortunate. Of the 41,300 British servicemen who had one or more of their limbs amputated as the result of war service, three-quarters lost a leg.[8] However, very few men endured an amputation as high as Hopkinson's. Of all the amputees who were treated (like Hopkinson) at Queen Mary's Hospital at Roehampton in London, only 11 (1.5 per cent) were left with such a short stump – that is, a stump that did not exceed five inches in length measured from the tip of the great trochanter.[9] In order to facilitate an artificial limb, the ideal length for thigh amputations was 10–12 inches.[10] In October 1918, the army's Medical Board reported that there was 'no sufficient covering on [Hopkinson's] femur'. He was left with:

> merely a skin flap[,] conditions of which is poor and painful on pressure. His condition is permanent, as any operation to remedy this would practically leave no stump to assist in [word missing] of artificial limb. Instructed to proceed Home.[11]

This was even before the third re-amputation in 1927. As a result, the only artificial limb Hopkinson could be given was the 'tilting table' artificial limb, a notoriously heavy and difficult limb to wear.[12] In fact, he was never able to wear an artificial limb and spent half a century on crutches. The Ministry of Pensions judged him to be 80 per cent incapacitated and, after a period of recuperation in his father's vicarage in Whitburn, he went to live in London. The local woman to whom he was engaged broke off the engagement and later married the super-fit Captain Percy de Winton Kitcat of the Wellesley Nautical School, Blyth.[13] His nephews remember him as their 'one-legged uncle' who drove a large sports coupe, in which sat a long-term male 'friend'.[14]

Hopkinson's amputations and the collapse of his marriage plans were complicated by another ordeal: he suffered from shell shock. For Hopkinson, his painful stump and his psychiatric instability were insep-arable and unproblematic. As he explained in a letter to the Ministry of Pensions in 1919:

> I have been invalided from the service and immediately find that I am unable to take up employment on account of my bad ampu-tation ... I have had to have my stump reamputated and it will be some months before I can bear the pressure of an artificial limb. I can, therefore, only walk on crutches – *my stump is very short* and almost amounts to Disarticulation. I suffer also from nervousness, insomnia and impaired memory, having been a patient in Palace Green Shellshock Hospital for some months.[15]

Indeed, Hopkinson's psychological distress had been closely related to his wounding. While being evacuated to King Edward VII Hospital, he 'had to wait some hours under Railway arches during Air Raid'. The strain was too much: upon his arrival in London, he was reported to have 'developed confusion of thought with suspicions and hostil-ity'.[16] On 1 October 1917, only a week after having his leg amputated a second time, he was sent for treatment to Palace Green Hospital for shell-shocked officers. On his arrival, he was described as having an 'anxious expression ... He is confused and suspicious of his surround-ings, doubtful as to dates and times. Afraid of air raids and anxious to be evacuated'.[17] It took nearly five months for his doctors to report 'Mental condition now clear'.[18]

Phantom pain

Although the psychological effects of Hopkinson's war service persisted throughout his life, medical attention initially focused on his phan-tom limb and painful stump. In themselves, phantom limbs were not unusual. The phenomenon had first been described in 1551 by the great French military surgeon Ambroise Paré. In Paré's words:

> A most clear and manifest argument of this false and deceitfull [*sic*] sense appears after the amputation of the member; a long while after they will complain of the part which is cut away. Verily it is a thing wondrous strange and prodigious, and which will scarce be credited, unless by such as have seen with their eyes, and heard with their ears the Patients who have many months after the cutting away of the Leg, grievously complained that they yet felt exceeding great pain of that leg so cut off.[19]

Most famously, these post-amputation sensations were brought to public attention during the American Civil War when neurologist Silas Weir Mitchell coined the term 'phantom limb' and went on to provide the first modern clinical description of 'these hallucinations ... so vivid so strange'. He observed that:

> Nearly every man who loses a limb carried about with him a constant or inconstant phantom of the missing member, a sensory ghost of that much of himself, and sometimes a most inconvenient presence, faintly felt at times, but ready to be called up to his perception by a blow, a touch, or a change of wind.[20]

Indeed, phantom limbs were experienced by nearly all amputees (they were so common that amputees called them 'plimbs'),[21] but most of these sensations were 'not unpleasant' and were 'frequently even pleasant'.[22] In the words of one sufferer in 1945, they felt like someone was 'striking your funny-bone'.[23]

Painful phantom sensations were a completely different matter. They were much more rare. Most surgeons estimated that fewer than 16 per cent of amputees complained of painful phantoms, with the vast majority placing the percentage at closer to one per cent.[24] For that unlucky minority, phantoms were not simply painful, but torturous. This was certainly the case with Hopkinson, who described them as feeling like 'the foot was being crushed but ... at stump level only'.[25]

Specialists reported that once phantom pains had become 'established', they tended to 'persist': in one group of sufferers, only one-third of those who experienced moderate or severe phantom pain *ever* found relief subsequently.[26] As two experts remarked in 1945, painful phantoms posed a 'formidable therapeutic problem' and 'treatment of this troublesome symptom is difficult at best' since 'no single method is successful in all cases'.[27] Even at the end of the twentieth century – when there were at least 68 distinctive treatments available for painful phantom limbs – surveys of thousands of phantom limb pain sufferers revealed that fewer than one per cent reported any significant benefits from *any* of the therapies on offer. At most, eight per cent reported experiencing partial or temporary relief after being treated.[28]

It comes as no surprise, therefore, that Hopkinson was frustrated by the inability of his physicians to eradicate, or even ameliorate, his agony. In October 1937, for instance, he reported suffering from 'pains like electric shocks' from his stump, which were only partly 'relieved by aspirin & whiskies & sodas'.[29] An unidentified physician recommended that Hopkinson try 'Antikamnia' and he concluded his report with the words 'very temperamental!!!'.[30] Hopkinson was unhappy about this examination, writing a few months later to complain. 'Under no stretch of the imagination could what takes place be called an examination', he grumbled:

> The Medical Officer who happened to be on duty & was extremely sympathetic merely put his hands on my 'Stump' & said nothing could be done for the pains which are becoming more frequent & of considerable duration, but I could take 'Antikamnia' which has no effect. From this examination I note that the Ministry's Medical Officers are of the opinion that nothing can be done for the rest of my existence to relieve these pains.[31]

Being dismissed with advice to take Antikamnia must have been particularly galling for Hopkinson. Antikamnia was an extremely common 'over the counter' remedy.[32] Most physicians disparaged it on the grounds that 'if we are to believe the vendors, it relieves everything from flatulence to locomotor ataxia'.[33] It contained sodium bicarbonate, caffeine, citric acid and (the analgesic ingredient) acetanilide or antifebrin, which was known to have serious side-effects such as cyanosis.[34]

According to physicians at the time and to Hopkinson, phantom limb pain and stump pain were closely related. He complained of 'persistent local cutaneaus tenderness of the stump'[35] as well as 'sharp electric shooting pains' that made him 'shout out'.[36] However, as time went on and no cure was found, he increasingly had difficulty persuading his doctors that his suffering was 'real' – which for him *and* the doctors meant having a *physiological* origin. Instead, medical reports focused more and more on the alleged neurotic character of his symptoms. It did not help that his 'shell shock' had not been a direct consequence of combat, but had begun when he was subjected to an air raid on his way to hospital after landing in England. This showed a lack of soldierly self-control (after all, these bombs were intended to terrify *civilians* rather than soldiers) and was further evidence that he had a pre-existing mental weakness. From the 1930s, medical reports increasingly reiterated the view that he was of a 'marked neurasthenic type'.[37] He was 'a highly introspective type & very resentful'.[38] In the words of the Medical Board in January 1936, he was a 'man of sensitive temperament, not a good type', perhaps an allusion to his homosexuality.[39] On 17 July 1939, yet another medical examination catalogued his stump's 'lightning pains', but concluded that:

> This officer is highly neurotic & the lesion is in his mind & not the stump. Complains of having been badly treated & that he is insufficiently compensated by a 70% [in fact, it was 80 per cent] pension. Thinks it should be 100%.[40]

For the remaining decades of his life, Hopkinson was plagued by accusations that his suffering was not 'real'. He had a 'curious hypochondriacal almost paranoid personality' (1951),[41] was 'not uncooperative but ... [was] firmly wedded to his symptoms and the disability generally' (1951)[42] and exhibited a 'hypochondriac type of personality – rather an "old womanish" type – egotistical and so on. Does not work & has all the time in the world to think about himself & his disabilities' (1952).[43] At the very least, Hopkinson's 'psychoneurosis' was 'constitutional' and the pensions authorities were therefore not liable to offer him compensation.[44] When he complained about the state of his hands (after spending decades using crutches, his hands 'always throbbing and aching; also wrists' and they 'present[ed] a discoloured raw beef appearance, about 4 [inches] x 2 [inches]'), the Medical Officer concluded that the root cause of his disorders was his 'underling constitutional condition' and 'general nervous disposition'.[45] For Hopkinson, it must have seemed like he was caught in an impossible bind. Every complaint was either a pre-existing or 'constitutional' one: it was as if the war never happened.

The chief problem for Hopkinson was that he was confronting deeply embedded clinical beliefs about stump and phantom pains. From the late 1930s onwards, it was widely assumed within the medical literature that these pains were neurotic in nature. In the words of Atha Thomas and Chester C. Haddan's important 1945 textbook, 'psychic factors' played such an 'important role' in painful phantoms that 'some authors are of the opinion that the phantom limb phenomenon represents some form of an obsessional neurosis'.[46] Authors of a 1947 article in *Psychosomatic Medicine* went so far as to claim that *all* patients complaining of phantom pain had 'severe psychopathology'.[47]

The hospital where Hopkinson received most of his treatment – Queen Mary's Hospital at Roehampton – was the leading centre for the scientific investigation of stump and phantom pains. In the 1950s, R.D. Langdale Kelham carried out a four-year study of 200 men with phantom limb pain. It is hard to imagine that Hopkinson would not have been one of these patients since he was exceedingly well known to all the doctors at Roehampton at this time. Kelham concluded that the typical phantom limb patient was:

> more often than not a person with an unsatisfactory personality. It may be he is an anxious, introspective, dissatisfied, ineffective [*sic*] who, becoming obsessed by his symptoms, and brooding upon them and his disability, tends to dramatise their degree, using undoubted exaggerations in his description of his sufferings.[48]

Kelham's assessment dominated the field. Only rarely did physicians suggest the opposite causality – in other words, that chronic pain might *lead to* psychological distress rather than being *caused by* it.[49]

A great deal was at stake. As we shall see shortly, one of Hopkinson's physicians recommended neurosurgery, but Hopkinson does not seem to have been offered the two other radical treatments proffered by those who believed that phantom pains were primarily the result of psychiatric shortcomings: electroconvulsive therapy (ECT) and frontal lobotomy. J.E. Pisetsky was one of many surgeons who advocated ECT for phantom pains. In 1946, he admitted that its effectiveness might be due to a host of factors, including:

> changes in the oxygen-carbon dioxide ratio, changes in the vascularity of various brain areas, changes in the blood pressure, velocity of blood flow, changes in the chemical contents of the blood, and changes in the cellular structure of the cerebral cortex.

However, Pisetsky also speculated that the painful treatment might work simply by 'jar[ring] the apathy or inertia which prevents the individual from facing reality'.[50] Given the frequent references in Hopkinson's medical files to his self-centredness, he might have been considered to be a candidate for these radical treatments. However, lobotomy for phantom limbs was much more common in America than Britain and Hopkinson was lucky to be treated by Dr Leon Gillis – the main physician at Roehampton – who happened to be opposed to such radical interventions. As Gillis concluded in his widely-admired textbook *Amputations* (1954):

> it is doubtful if such procedures are justifiable, as they do not in themselves abolish the perception of pain, but merely produce changes in the personality of the individual. The change is mostly one of lessening drive in initiative.

Gillis also disparaged such surgery on the grounds that it 'dull[s] the patient's mind'.[51] This should not be taken to mean that Gillis did not believe that stump and phantom pains were primarily psychiatric problems. He accepted that patients who were lower on the 'phylogenetic scale' or were 'highly intelligent, sensitive' were more at risk of developing phantom pains compared to 'more plethoric, unimaginative individuals'. However, he warned that 'fear of accusation of insanity' made the amputee:

> reluctant to talk about his symptoms and he prefers to hide them until he can no longer bear them. This may result in gross mental disturbance, and the sufferer of a phantom limb is often regarded as psychotic.

In other words, Gillis was one of those rare physicians who believed that long-endured suffering could *cause* psychiatric problems as well as being a sign of such disorders. He advised physicians to reassure amputees that, despite having a psychological etiology, their pains were 'very real' and would 'eventually' fade.[52]

Aside from reassurance (which even Gillis acknowledged lost its efficacy for men like Hopkinson, who had experienced decades of continual suffering), what therapies were offered to Hopkinson? In the four decades between the 1940s and his death, he was treated by physicians on both sides of a major divide in the treatment of stump pain – in shorthand, this was the difference between those who focused on the brain's reaction to painful stimuli ('centralists' or cerebralists) and those who were peripheralists (the painful sensations originated from 'excitation of nerve ends' in the scar or stump).

Cerebral theorists placed their bets on the efficacy of neurosurgery. On 23 November 1943, Geoffrey Jefferson, the 'doyen of neurosurgeons' who had conducted significant research (albeit unpublished) on phantom limb pain during the First World War and into the 1920s, examined Hopkinson.[53] He reported that Hopkinson experienced phantom pains 'really badly for a few hours every two months or so. It is then almost unbearable, he takes dope and it comes under control'.

Jefferson offered Hopkinson an alternative, radical treatment: chordotomy (also spelt cordotomy),[54] a treatment winning many medical converts from the 1940s. Chordotomy involved dividing the pain pathways in the spinal cord, thus 'interrupting the pathways of the painful impulses in order to abolish or modify their effects on the sensorium, either before they reach it or in the brain itself'.[55] Many neurosurgeons believed it could reduce or even eliminate intractable pain.

A particularly eloquent defence of chordotomy was mounted by Murray A. Falconer, the Director of Guy's Maudsley Neurosurgical Unit. For Falconer, the effectiveness of the treatment was itself proof that phantom limb pain was not 'a psychological disturbance'. Falconer did admit that, prior to the operation, some of his patients were 'greatly demoralized by pain, and were perhaps unstable individuals', but, he insisted, they were no more disturbed than patients who suffered from 'other organic painful conditions, such as trigeminal neuralgia'. Indeed, he continued, psychiatrists had been able neither to 'find any significant psychogenic features' in phantom limb patients nor to relieve suffering by psychotherapy. As for any suggestion that performing a major operation might itself be curative for *psychological* reasons, he was dismissive: 'I find it difficult to believe', he scoffed, 'that in my hands antero-lateral chordotomy acted as a psycho-therapeutic procedure, when previous operative procedures on the stump had failed to give relief.'[56]

However, the effectiveness of this surgery in alleviating the pain of phantom limbs had its critics. The operation required surgeons to guess how much of the spinothalmic tract in the spinal cord should be cut to maximise pain relief without disrupting other functions. In 1943 (that is, three years after Jefferson had offered the operation to Hopkinson), Jefferson had been persuaded to perform a chordotomy on a fellow surgeon, without success. As Jefferson's biographer reported, 'chronic pain is not usually relieved by this means. Jefferson must have known this and had operated against his better judgement'.[57] Perhaps the biographer was benefiting from hindsight. After all, Jefferson had clearly been carrying out the operation on other patients: in 1952, he reported mixed results in the chordotomies that he had performed on 12 sufferers of phantom limbs.[58] As critics of the operation pointed out, chordotomies often led to distressing side-effects, including 'defective sphincter control', 'motor defect' and 'decubitus ulcers'; it should not be 'lightly undertaken in the chronically ill and elderly'.[59]

Jefferson did not say why Hopkinson 'does not desire an operation' (perhaps the chronically weakened 54 year old was warned of these rather daunting risks) and Jefferson was also willing to 'leave matters as they are' with Hopkinson.[60] After all, although Jefferson was a passionate neurosurgeon, he claimed that he never believed that 'operation is the only method of treatment ... There are surgeons who think so, but I was never one of them. I hold that man a bad doctor who has but one method of treatment'.[61] Once Hopkinson convinced him that 'he was in no danger of becoming a drug addict', Jefferson sent him for more traditional forms of treatment.[62]

However, after refusing Jefferson's offer of a chordotomy, Hopkinson's misery simply got worse. Even the Ministry of Pensions – which had been refusing his claim for increased compensation for decades – finally admitted that his war-related disabilities were deteriorating. In March 1949, the Ministry reassessed his medical state, upgrading his percentage of disability from 80 per cent to 90 per cent. They noted that he lived in constant distress, experienced phantom pains, did not have an artificial limb and was totally dependent on elbow-crutches to move.[63]

Clearly other treatments had to be attempted. This time, rather than turning to the knife, Hopkinson's doctors tried physical medicine. The first choice – prescribed in 1939 and 1949 – was anodal galvanism.[64] Although often used to alleviate disorders as different as trench foot, neuralgia, neuritis, lumbago and sciatica,[65] anodal galvanism was also prescribed for phantom pain (as P. Jenner Verrall, one of Hopkinson's physicians, attested in the *British Medical Journal*).[66] It worked primarily as a sedative. As one advocate explained, the doctor placed a positive electrode at 'the extremity of the affected limb with the negative [electrode] towards the nerve root'. Its sedative effect was:

explained by the hyperaemia induced by the current being more pronounced near the anode or positive pole. As sensory nerve endings are stimulated by chemical or physico-chemical changes, pain produced by the arrest or diminution of the blood flow may be relieved by any measure that increases the circulation in the part concerned.[67]

In Hopkinson's case, anodal galvanism did help – but only temporarily.[68]

By 1951, another physical, or 'peripheralist', remedy was being attempted. Percussion therapy was the brainchild of neurologist W. Ritchie Russell, then based at the United Oxford Hospitals. Russell had a formidable reputation for his research on intractable pain. He claimed that 'percussion' was an effective way to banish stump pain. Just two years before he treated Hopkinson, he had shown a group of expert specialists on intractable pain a film illustrating his method of treating painful amputation stumps by repeated 'percussion' of the neuroma (tumours of nervous system tissue) with a mallet and wooden applicator.[69] As a physician who worked at Roehampton while Hopkinson was being treated explained, the mallet used was an 'ordinary, wooden carpenter's type' of about 1–1½ lbs in weight while the wooden peg was constructed from a broom handle or end of a crutch and was approximately six inches long. The peg should:

have fixed to one end a circular piece of metal, that commonly used for the legs of chairs, etc., known as the 'dome of silence' serves very well – and to the other end a crutch or walking stick rubber. The smooth metal covered end of the peg is applied firmly to the palpable neuroma, tender area or scar ... and the other rubber covered end is struck with the mallet as vigorously and as rapidly as is possible without giving the patient undue discomfort.[70]

In a different version, an electric vibrator was used in a similar way. Once the amputee's stump or phantom pains had begun to subside – which generally took between 24 and 48 hours of treatment, although it could take a month – he would be taught how to do it himself at home.[71]

How did percussion work? In 'Painful Amputation Stumps and Phantom Limbs: Treatment by Repeated Percussion to the Stump Neuromata' (1949), Russell explained that 'treatment at the periphery' (as opposed to the centre, as with Jefferson's surgery) was likely to be effective for three reasons: first, he observed that, even in normal limbs, nerve endings would be 'rendered insensitive by occupations which involve repeated minor trauma or prolonged firm pressure on the skin'; second, 'conduction of a mixed nerve is easily interrupted by repeated pressure', without causing pain; and, third, 'the regenerating nerve fibres in an amputation stump are likely to be no less vulnerable to minor trauma or pressure than are normal nerves and nerve endings'. Russell recommended that an amputee should carry out this procedure twice a day – he would, literally, be 'learning to knock away his phantom pain whenever it was troublesome'.[72] It was, the *British Medical Journal* pronounced in 1949, a 'refreshingly simple method'.[73] Indeed, it is still used by some stump pain sufferers today.[74]

Although a disarmingly simple procedure, proponents acknowledged that it was important that percussion treatment was initially carried out within a hospital ward. As a rehabilitation physician at Roehampton remarked in the 1950s, the rationale of percussion treatment had to be explained to the patient, since 'some are inclined to be a little intimidated by the prospect of their painful stumps being assaulted by what at first appears to them to be a somewhat violent procedure'.[75]

Russell reported that Hopkinson had benefited from percussion treatment when he had been admitted to Queen Mary's Hospital at Roehampton. While in the ward, Hopkinson experienced 'much less sensitivity in the stump than before, and has had no severe bouts of pain since leaving'. However, Russell warned that there was a serious danger of a relapse, since Hopkinson had subsequently 'discontinued trying to treat himself'. The fact that he had been 'supplied with an applicator but no mallet' was not encouraging. Russell recommended that Hopkinson should be urged to 'persevere with self-administered percussion treatment', even though he would have to buy the mallet with his own money.[76] Indeed, R.D. Langdale Kelham (who had concluded his study of 200 men with phantom limb pain at Roehampton three years earlier) had someone remarkably similar to Hopkinson in mind when he attempted to explain the high relapse rate in phantom pains after the amputees left the hospital. Although percussion treatment had been effective in the wards, Kelham observed, a follow-up study revealed that only 30 per cent had been able to control their pain at home. Kelham ascribed this high relapse rate to the fact that:

Such cases often have unstable personalities, are often paranoid in type, their experiences in the past have been discouraging and have only served to strengthen the conviction that they have something seriously wrong in their stumps and that nothing can be done to help them. They do not view new methods of treatment with any optimism and their whole attitude becomes negative and defeatist ... As soon as they are removed from the influence of in-patient conditions with its constant encouragement, they were foredoomed to relapse, because, like some of the initial failures, they were unsuitable material in the first place.[77]

Whether Hopkinson's problem could be blamed on his physiology or psyche, percussion treatment eventually failed. Five years after his consultation with Russell, a medical report concluded that mallet and peg treatment had 'little lasting effect' on his stump pain.[78] Although he reported that the *electric* vibrator version of the mallet and wooden applicator did help relieve some pain, his physicians were increasingly frustrated by their inability to completely alleviate it. Comments on his poor psychological adaptation began appearing more frequently. In 1958, for example, a medical report stated that Hopkinson was still experiencing a 'jumpy stump ... Finds holding it helps ... psychopathic personality'. He quoted Hopkinson as saying 'I am frightened of the pain. I do not mind making scenes'.[79] Or, as another doctor noted a month later, Hopkinson 'probably does get pain', but it was '100% aggravated by his poor mental adaptation & aggravated by low pain tolerance'. In an exasperated tone, he concluded: 'One wonders whether if it is worth bothering with him but ... he is firmly convinced that vibrators do help him – even if it is a mental placebo.'[80] This was the kind of cynical 'complacency' that the great pain surgeon René Leriche might have been alluding to when, in 1939, he wrote about the 'rather bizarre geography of subjective symptoms' that men with painful stumps described. 'As is usually the case when we fail to understand anything', he thoughtfully concluded, 'we ascribe an important part to imagination and emotion.'[81]

Ageing amputee

Hopkinson's predicament was representative of a wider trend in British society after the Second World War. All the physicians who treated him were aware that Hopkinson was part of a much bigger problem associated with 'elderly amputees'. In 1953, it was estimated that there were nearly 24,000 men (and one woman) in Britain who had lost one or more limbs as a result of war injuries that had occurred during the First World War or before.[82] Amputees from the First World War were now elderly and the newly established National Health Service was struggling to deal with them.

Awareness of this crisis led Donald Stewart McKenzie – who had examined Hopkinson in the 1940s – to conduct research on this constituency of disabled men. It was eventually published as 'The Elderly Amputee' in 1953. McKenzie began by noting that techniques for rehabilitating amputees had 'evolved primarily in relation to active ex-Service men' – in other words, they had been devised to restore young, fit men. What physicians and pension authorities were facing by the 1950s, though, was a more 'enfeebled' set of patients for whom previous approaches simply could not be applied. McKenzie placed some of the blame for the lack of progress with this new generation of elderly amputees on 'unwarranted optimism'. Patients had been taught to *expect* that they would be able to walk with a prosthesis 'without effort' and were consequently demoralised when faced with the magnitude of the challenges facing them.[83] More importantly, however, McKenzie emphasised the importance of 'environment'. In his words:

> We not infrequently see patients who have made good progress and who are discharged from the walking school able to control their prosthesis and to look after themselves fairly well. Yet when we see them on follow-up we find they have hardly worn the prosthesis, and the musculature has lost tone to the extent that they can no longer control it. Inquiry reveals that they live alone, perhaps in an upstairs flat, or it may be that they simply lacked the incentive to make the effort to persevere.[84]

McKenzie's description closely matched Hopkinson's circumstances. After all, Hopkinson had been able to control his pain effectively using the 'percussion' method while in hospital, but lapsed when he returned home. He was never able to wear an artificial limb. By the time McKenzie was writing, Hopkinson perfectly fitted McKenzie's profile of the 'elderly amputee' who complained of chronic stump and phantom pain, shell shock and was reaching an age when he was finding it difficult to clean himself and his flat.[85]

However, two final attempts were made to treat Hopkinson's stump pain. On 17 October 1956, Leon Gillis, the limb expert who had published his highly influential book *Amputations* two years earlier, took on Hopkinson's case. In *Amputations*, Gillis had warned that:

> Pain is a symptom, and however important it is to the patient to be given relief from his pain, it is more important still that the cause of a painful stump should be accurately determined before any treatment is started.

Gillis strongly believed that treatment needed to encompass 'psychological as well as physiological factors' and while he conceded that the 'psychological element may play an insignificant part … on the other hand it may be the sole cause of the pain', particularly in 'a world which is fraught with economic crises, social maladjustment, anxiety, and fear'. These external influences acted as 'powerful factor[s] in increasing the perception' of stump pain.[86]

Gillis immediately observed that there was an emotional element to Hopkinson's suffering. He reported that Hopkinson 'thinks the pain is affected by change in weather and also "when he gets annoyed"'. The only way forward was to get a full range of pain and limb specialists together to investigate what had been going wrong for 39 years. In October 1956, Gillis enlisted the help of the Painful Stump Panel.[87]

Hopkinson was admitted to Queen Mary's Hospital where he was to remain for a month. In the Panel's report, they observed that Hopkinson was a 'big man' with a very short stump. An x-ray of his stump showed 'some bony spurs', often believed to cause pain. Each of the specialists gave their diagnosis and each responded predictably according to their specialism. Psychiatrist Guy Randall, who had co-authored an article entitled 'Psychiatric Reactions to Amputation' (1945), believed that 'there was a considerable psychiatric factor in this case'. He noted a 'family background of instability' and claimed that Hopkinson's irregular employment record reflected 'personal instability'. The only solutions, Randall claimed, were 'sedative therapies, e.g., Equanil, Largactil or Phenergan alone or with barbiturates' since there was 'little or no chance of altering his personality or reaction type'.[88] Consultant neurologist Dr Aldren Turner recommended that Hopkinson return to percussion treatment, supplemented with analgesics 'during severe attacks', while consultant orthopaedic surgeon Mr Harding 'wondered whether his prostatectomy in 1951 may have been related'. The summary ended by noting that Hopkinson had undergone percussion therapy and was taking the mild tranquilisers Equanil (meprobamate) and Sonalgin, which contained the sedative butobarbitone and was advertised as being 'valuable for the relief of nervous tension'.[89] The medication seemed to work. The Panel reported that Hopkinson 'now feels able to cope with life again and requests discharge'.[90]

Clearly, the 1956 prognosis was optimistic: within two years, Hopkinson was yet again experiencing severe pain in his stump and had serious mobility problems.[91] In March 1964, Dr Ian H.M. Curwen, Consultant on Physical Medicine, took over Hopkinson's case. Curwen reported that 75-year-old Hopkinson 'experiences pain for one or two days about every four weeks ... He says that his prostate has been "partly removed" ... the whole of his spine moved poorly and he probably has gross spondylosis', or degenerative osteoarthritis.[92]

Once again, Hopkinson was admitted into the rehabilitation ward in Roehampton. This time, a different peripheralist treatment was tried. Between 22 April and 11 May 1964, Lignocaine (a common anaesthetic, otherwise known as Lidocaine and Xylocaine) was injected into his stump daily. The medical record observed that: 'Immediately following injection he felt a pleasant warm sensation in the stump. Nocturnal discomfort in the stump was reduced and he had one incident only of stump pain during his two weeks' admission. This was less severe than usual and lasted only a few hours.'[93] In July 1964, Hopkinson was still improving. He 'gets a little stump pain [but] never severe. Minutes only now and never hours. Has less "rawness" in his short phantom'. The injections seemed to be effective.[94]

The reason for the success of Xylocaine injections was disputed. Hopkinson's former doctor – W. Richie Russell, the percussion specialist – explained that when dealing with patients with intractable pain, it was not necessary to know precisely *why* a certain treatment worked, so long as it did. He reminded his fellow doctors that it made 'no sense saying that one pain is functional and one organic' because 'all pains are both physiologically determined and functionally graded according to a wide variety of personal factors'. Even when the pain was largely the result of emotional factors, physical treatments might work. In the case of percussion treatment, if the 'discharging neuromata' were 'inactivated', the patient would report some alleviation of his pain even though emotional responses meant that the pain would never be totally eradicated. Similarly, Russell continued, Xylocaine injections given into the 'anatomical area concerned in some way with a chronic pain' might help to break the 'vicious cycle' by providing temporary relief. In this way, the local anaesthetic would have a 'curative as well as a diagnostic value'. This was why he was even willing to endorse 'old methods of treating pain with electricity' (that is, the galvanism treatment Hopkinson received in 1949) since, at the very least, it would provide 'a physiological distraction' that might actually reduce suffering. As Russell wittily contended at the very end of his paper: 'I would suggest that the successful therapist for intractable pain treats the problem like a game in which he endeavours to outmanœuvre the tricks played by the C.N.S. [central nervous system] of his patient.' The therapist:

> had many different moves he can play. Some depend on simple procedures which checkmate the mechanisms, but others are assisted by the deception of the poker player and the confidence of the quack. I may add that my colleagues ... think that I am too optimistic about the results of treatment, but I think it important to be over-confident in treating pain, so I make no apology.[95]

Russell believed in the power of mind over the body: the problem in Hopkinson's case was that he was too disillusioned and too disenchanted to believe in any positive outcome, let alone the intentions of physicians working for any governmental ministry.

It took another three years before the Medical Board finally accepted what Hopkinson had been telling them all along. At the age of 84 and 56 years after he had been wounded in the war, they accepted that he really was 100 per cent disabled. He was judged to be 80 per cent disabled because of his amputated leg, 6–14 per cent disabled due to the injury to his left elbow and 10 per cent disabled because of 'foreign bodies' (that is, bomb fragments) in his left shoulder, knuckles and wrist, the presence of which his physicians had always denied. They also reported that he had a head injury, osteoarthritis in his right wrist and thumb, and 'callosity' in the palm of his hand due, no doubt, to more than half a century on crutches.[96] He was offered physiotherapy.[97] A report by the Ministry of Social Security on 28 January 1974 noted that Hopkinson was 'depressed at times from pain' and he was experiencing '*Severe phantom* pains by day with stabs of stump pain'. After a 'friend' from the British Limbless Ex-Service Men's Association offered to take him to Brighton for a holiday if the Ministry would pay the cost of petrol,[98] Hopkinson pleaded with the Welfare Officer to allow him to go because 'I am tired of sitting alone in my Bedroom ... except for 2 hours outing on Sundays & it is bad for my morale'.[99] Hopkinson died on 17 December 1974. He was 85 years of age and had lived for 57 years with war injuries. Under cause of death, the death certificate recorded: 'SENILITY, MYOCARDIAL DEGENERATION AND FAILURE.'

Conclusion

From the age of 28 until the age of 85, Frank Hopkinson had lived in almost constant pain. One of his doctors had reported that he:

gets a lot of twitching & jumping in the stump – like electric shocks – makes him shout & gets a temperature. Comes on at irregular times – about 4 times a year – often with a change in the weather – or if he goes to stay with a friend. Emotional – cries if he hears a hymn or if he can't get a seat in a bus. Very irritable; cannot concentrate; unreliable. Sleep good. Single. Has tried to get a job, but always turned down.[100]

Although Hopkinson's symptoms changed relatively little throughout his life, his sufferings cannot be summarised under any single headings. His pain was acute, chronic, physiological, psychological and emotional; it gripped him within hospital wards and when he was 'sitting alone in my Bedroom'. He struggled to distinguish the experience of pain from the pain of experience. On the surface, he should have been able to elicit sympathy: he was a white male who had been born into a privileged family and had served as an officer in war. In fact, his class status was a further cause for agony. As one doctor reported: 'The officer is a man of sensitive temperament and a loss of his leg affects him more than one of coarser fibre. He ... hates people looking at him and sympathizing with him.'[101] Those physicians who witnessed his pain often attempted to sympathise and provide succour, but their inability to solve his crises eventually led each of them to turn away – sometimes in despair, at other times in annoyance. The invisibility of his wound – his stump seemed to be 'normal' and the limb that burned like fire did not exist – trumped all scientific theorising. Theories about physiological pain pathways, psychiatric pathologies, constitutional inheritances, psychosomatic symptoms and even 'old womanish' sensitivities failed to ease suffering that was anything but 'phantom'.

Notes

1. I am immensely grateful to the Wellcome Trust for its generous financial support in setting up the Birkbeck Pain Project (BPP). The research would not have been possible without the intellectual support and inspiration of my two colleagues in the BPP, Dr Louise Hide and Dr Carmen Mangion.
2. Filippo de Vivo, 'Prospect or Refuge? Microhistory, History on the Large Scale. A Response', *Cultural and Social History*, 7(3) (2010): 387. For good introductions, see Peter Burke, *New Perspectives in History Writing* (University Park: Pennsylvania University Press, 1991); and C. Ginzburg, 'Microhistory: Two or Three Things I Know About it', *Critical Inquiry*, 201 (1993): 10–35.
3. Seth Koven, 'Remembering and Dismemberment: Crippled Children, Wounded Soldiers, and the Great War in Great Britain', *American Historical Review*, 99(4) (1994): 1169.
4. Interview of Giles and Ben Hopkinson, 6 November 2012.
5. 'Report of Medical Board', National Archives WO 339/12060 (P1030322) and 'Special Hospitals for Officers', 7 March 1924, National Archives, PIN 26/21799.
6. Medical reports, National Archives, PIN 26/21799 Part 2 (P1040118–P1040119).
7. Letter from Hopkinson, 4 December 1918, National Archives WO 339/12060.
8. E. Muirhead Little, *Artificial Limbs and Amputation Stumps. A Practical Handbook* (London: H.K. Lewis and Co., 1922), 23. The percentage was 73 per cent. For a longer discussion of war wounds, see Joanna Bourke, *Dismembering the Male: Men's Bodies, Britain, and the Great War* (London: Reaktion, 1996).
9. Muirhead Little, *Artificial Limbs*, 24.

10. A.W.J. Craft, 'Amputations, Limb Fitting, and Artificial Limbs. Lecture Delivered at the Royal College of Surgeons of England on 11th April, 1949', *Annals of the Royal College of Surgeons of England*, 5(3) (1949): 194. See also P. Jenner Verrall, 'President's Address. Some Amputation Problems', *Proceedings of the Royal Society of Medicine*, 24(2) (1930): 183.

11. 'Proceedings of a Medical Board... Lieut. F. Hopkinson' (14 October 1918), National Archives, PIN 26/21799.

12. See R.D. Langdale Kelham, *Artificial Limbs in the Rehabilitation of the Disabled* (London: Her Majesty's Stationery Office, 1957), 82; Atha Thomas and Chester C. Haddan, *Amputation Prosthesis. Anatomic and Physiologic Considerations, with Principles of Alignment and Fitting Designed for the Surgeon and Limb Manufacturer* (Philadelphia: J.B. Lippincott Company, 1945), 115; 'The Weight of Limbs: Natural and Artificial Limbs Compared', *British Medical Journal* (24 August 1918), 202. See also Jensen J. Steen and T. Mandrup-Poulsen, 'Success Rate of Prosthetic Fitting after Major Amputations of the Lower Limb', *Prosthetics and Orthotics International*, 7 (1983): 119–21; and D.G. Shaw, T.M. Cook, J.A. Buckwalter and R.R. Cooper, 'Hip Disarticulation: A Prosthetic Follow-Up', *Prosthetics and Orthotics International*, 37 (1983): 50–7.

13. 'Forthcoming Marriages', *The Times*, 15 January 1919, 11; and 'Forthcoming Marriage', *The Times*, 20 November 1928, 19.

14. Interview with Giles and Ben Hopkinson, 6 November 2012.

15. Letter from Hopkinson to the Minister of Pensions, 1 January 1919, National Archives, PIN 26/21799, emphasis in original.

16. 'Special Hospitals for Officers', 7 March 1924, National Archives, PIN 26/21799 and Report in National Archives PIN 26/21799 part 2 (P1040144). See also the report dated 12 August 1917, National Archives PIN 26/21799 part 2 (P1040144).

17. 'Medical Care Sheet', National Archives PIN 26/21799 part 2 (P1040121-P1040122).

18. *Ibid.* He was dismissed from Palace Green on 19 February 1918.

19. Ambrose Paré, quoted in Thomas Johnson, *The Works of that Famous Chirurgion, Ambrose Parey [sic], Translated Out of Latine [sic] and Compared with the French* (London: n.p., 1649). See also Ambrose Paré, *Oeuvres Complètes* (ed. J.F. Malgaigne), vol. 2 (Paris: 1840), 221–31.

20. S. Weir Mitchell, *Injuries of Nerves and their Consequences* (Philadelphia: J.B. Lippincott and Co., 1872), 348.

21. Lieutenant Colonel Guy C. Randall, Jack R. Ewalt and Lieutenant Colonel Harry Blair, 'Psychiatric Reactions to Amputation', *Journal of the American Medical Association*, 128(9) (1945): 645–52.

22. See W.R. Henderson and G.E. Smyth, 'Phantom Limbs', *Journal of Neurology, Neurosurgery, and Psychiatry*, 11 (1948): 89–92; and Bertram Feinstein, James C. Luce and John N.K. Langton, 'The Influence of Phantom Limbs', in Paul E. Klopsteg and Philip D. Wilson (eds), *Human Limbs and their Substitutes. Presenting Results of Engineering and Medical Studies of the Human Extremities and Application of Data to the Design and Fitting of Artificial Limbs and to the Care and Training of Amputees* (New York: McGraw-Hill, 1954), 80–1. See also John C. Wellons, John P. Gorecki and Allan H. Friedman, 'Stump, Phantom, and Avulsion Pain', in Kim Burchiel (ed.), *Surgical Management of Pain*

(New York: Thieme, 2002), 427; 'Phantom Limbs', *Journal of the American Medical Association*, 125(9) (1944): 633–4.

23. 'Phantom Limb Syndrome', *Journal of the American Medical Association*, 128(12) (1945): 904.

24. J. Lawrence Pool, 'Posterior Cordotomy for Relief of Phantom Limb Pain', *Annals of Surgery* (1946): 386; Murray A. Falconer, 'Surgical Treatment of Intractable Phantom-Limb Pain', *British Medical Journal* (1953): 299–304; Wellons *et al.*, 'Stump, Phantom, and Avulsion Pain', 424; Kelham, *Artificial Limbs*, 137.

25. Unnamed Medical Officer, Ministry of Pensions, Medical Case Sheet, 6 November 1951, National Archives, PIN 26/21799.

26. Bertram Feinstein, James C. Luce and John N. K. Langton, 'The Influence of Phantom Limbs', in Klopsteg and Wilson (eds), *Human Limbs and their Substitutes*, 85.

27. Thomas and Haddan, *Amputation Prosthesis*, 59–60.

28. Richard A. Sherman, 'History of Treatment Attempts', in Richard A. Sherman (ed.), *Phantom Pain* (New York: Plenum Press, 1997), 143.

29. Unsigned medical report for the Ministry of Pensions, 5 October 1937, National Archives, PIN 26/21799.

30. *Ibid.*

31. Letter from Hopkinson to the Secretary of the Ministry of Pensions, dated 1 January 1938, National Archives, PIN 26/21799. For the physician's response, see his letter dated 10 January 1938, National Archives, PIN 26/21799 part 2 (P1030659).

32. '"Patent Medicines" in New York', *British Medical Journal* (1915): 601–2.

33. 'The Nostrum Nuisance', *British Medical Journal* (1910): 1073–4.

34. 'Patent Medicines', *British Medical Journal* (1903): 1654; and 'The Nostrum Nuisance', *British Medical Journal* (1910): 1073–4. Cyanosis is dangerously low levels of oxygen in tissues near the skin surface.

35. Murchell in 'Ministry of Pensions Medical Case Sheet', 17 July 1938, National Archives, PIN 26/21799.

36. Unnamed Medical Officer, report to the Ministry of Pensions, 3 October 1939, National Archives, PIN 26/21799. None of these physicians disputed the phantom pains, simply their cause: see also Unnamed Medical Officer, report to Ministry of Pensions, 13 October 1939, National Archives, PIN 26/21799.

37. Reported in a letter from F. Murchie of the Director General of Medical Services, 20 December 1935, National Archives, PIN 26/21799. See also memo dated 12 August 1932, National Archives PIN 26/21799 part 2 (P1030703); memo 1 August 1932, National Archives PIN 26/21799 part 2 (P1040099); 16 January 1936, National Archives PIN 26/21799 part 2 (P1030705).

38. See report of 3 February 1936, National Archives PIN 26/21799 part 2 (P1040090).

39. Medical Board report, 2 January 1936, National Archives, PIN 26/21799 part 2 (P1030636).

40. Unnamed Medical Officer, Ministry of Pensions, 19 July 1939, National Archives, PIN 26/21799 part 2 (P1030606).

41. Unnamed Medical Officer, Ministry of Pensions, Medical Case Sheet, 6 November 1951, National Archives, PIN 26/21799.

42. Unnamed Medical Officer, Ministry of Pensions, Medical Case Sheet, 6 November 1951, National Archives, PIN 26/21799.
43. Unsigned medical report, 1 December 1952, National Archives, PIN 26/21799. For physicians who made this argument, see Falconer, 'Surgical Treatment', 301.
44. Report dated 28 August 1954, National Archives PIN 26/21799 part 2 (P1040086).
45. Report by Hopkinson, 7 February 1940, National Archives, PIN 26/21799 part 2 (P1030597) and Report by Chief Regional Officer, dates 8 February 1940, National Archives, PIN 26/21799 part 2 (P1030595).
46. Thomas and Haddan, *Amputation Prosthesis*, 59.
47. J.R. Ewalt, G.C. Randall and H.D. Morris, 'The Phantom Limb', *Psychosomatic Medicine*, 9 (1947): 118–23.
48. Kelham, *Artificial Limbs*, 131 and 139.
49. W.F. Tissington Tatlow and J.L. Oulton, 'Phantom Limbs (with Observations on Brachial Plexus Block)', *Canadian Medical Association Journal*, 73 (1955): 173.
50. J.E. Pisetsky, 'Disappearance of Painful Phantom Limbs after Electric Shock Treatment', *American Journal of Psychiatry*, 102 (1946): 599–60. For other advocates of ECT for phantom pains, see 'Electric Shock in Painful Phantom Limb', *Journal of the American Medical Association*, 131(11) (1946): 942; and Lothar B. Kalinowsky and Paul H. Hoch, *Shock Treatments, Psychosurgery, and Other Somatic Treatments in Psychiatry*, 2nd revised edn (New York: Grune and Stratton, 1952), 204.
51. Leon Gillis, *Amputations* (London: William Heinemann Medical Books, 1954), 266.
52. Leon Gillis, 'Pain in Phantom Limbs', *British Medical Journal* (1948): 1108.
53. 'Doyen of Neurosurgeons', *British Medical Journal* (1960): 1788; and Peter H. Schurr, *So That Was Life: A Biography of Sir Geoffrey Jackson* (London: Royal Society of Medicine Press, 1997), 114 and 131.
54. The original spelling was 'chorotomy', but 'cordotomy' (the American spelling) has become more common. As a Greek word, 'ch' is a more accurate transliteration.
55. Lambert Rogers, 'Refresher Course for General Practitioners: The Surgical Relief of Pain', *British Medical Journal* (1952): 383.
56. Murray A. Falconer, 'Surgical Treatment if Intractable Phantom-Limb Pain', *British Medical Journal* (1953): 301. For another positive assessment of its effectiveness, see 'Chordotomy for Painful Phantom Limb', *Journal of the American Medical Association*, 132(2) (1946): 112; and J. Lawrence Pool, 'Posterior Cordotomy for Relief of Phantom Limb Pain', *Annals of Surgery* (1946): 390.
57. Schurr, *So That Was Life*, 255.
58. Geoffrey Jefferson speaking at 'Second Plenary Session: The Relief of Pain', *British Medical Journal* (1952): 147.
59. J. Donaldson Craig, 'Pain in Phantom Limbs', *British Medical Journal* (1948): 904; J.D. Parkers, 'Diseases of the Central Nervous System', *British Medical Journal* (1975): 92; Rogers, 'Refresher Course', 383; P. Jenner Verrall, 'War Surgery of the Extremities: Amputations', *British Medical Journal* (1942): 677; 'Intractable Pain', *British Medical Journal* (1968), 513–14; 'Surgical Relief of

Intractable Pain', *British Medical Journal* (1961): 663–4; John C. Brocklehurst, 'Old Folks in Wet Beds', *British Medical Journal* (1962): 115. For other negative assessments, see J.A.W. Bingham, 'Pain in Phantom Limbs', *British Medical Journal* (1948): 52; Gillis, *Amputations*, 266; Kelham, *Artificial Limbs*, 143; Thomas and Haddan, *Amputation Prosthesis*, 60; Wellons *et al*., 'Stump, Phantom, and Avulsion Pain'; James C. White, 'The Problems of the Painful Scar', *Annals of Surgery*, 148(3) (1958): 422–31.

60. Geoffrey Jefferson, 'Report of Specialist', 24 November 1943, National Archives, PIN 26/21799. Jefferson's address was 3 Lorne Street, Manchester.

61. Geoffrey Jefferson, 'Treatment of Trigeminal Neuralgia', *British Medical Journal* (1932): 223.

62. There had been concerns about Hopkinson's drug habit in 1932 and 1936. See National Archives PIN 26/21799 part 2 (P1030703), (P1040099) and (P1030705). The fear that chronic sufferers would 'become narcotic addicts' was frequently repeated: see Thomas and Haddan, *Amputation Prosthesis*, 59.

63. Ministry of Pensions, 'Pensioner's Medical Record', 1 March 1949, National Archives, PIN 26/21799.

64. See report dated 13 October 1939, National Archives, PIN 26/21799 part 2 (P1030600)

65. 'Nitro-Glycerine in Trench Feet', *British Medical Journal* (1917): 513; and Matthew B. Ray, 'Physical Methods of Treating Rheumatism', *British Medical Journal* (1936): 1310.

66. Verrall, 'War Surgery of the Extremities', 677.

67. Ray, 'Physical Methods', 1310.

68. Medical Officer report to the Ministry of Pensions, 10 August 1949, National Archives, PIN 26/21799.

69. 'Intractable Pain', 84. See also W. Ritchie Russell, 'Painful Amputation Stumps and Phantom Limbs: Treatment by Repeated Percussion to the Stump Neuromata', *British Medical Journal* (1949): 1024. He was a neurologist who also wrote on facial palsy, poliomyelitis, head wounds, amnesia and multiple sclerosis. In 1975, he wrote *Explaining the Brain* (Oxford University Press).

70. Kelham, *Artificial Limbs*, 141.

71. *Ibid*.

72. Russell, 'Painful Amputation Stumps', 1024–6.

73. 'Pain in Phantom Limbs', *British Medical Journal* (1949): 1132.

74. Wellons *et al*., 'Stump, Phantom, and Avulsion Pain', 425.

75. Kelham, *Artificial Limbs*, 141.

76. From W. Ritchie Russell, to Dr R.D.L. Davies, 7 December 1951, National Archives, PIN 26/21799.

77. Kelham, *Artificial Limbs*, 143.

78. Note by R.D.L. Davies, dated 12 October 1956, National Archives, PIN 26/21799.

79. Dr Harbour, in Ministry of Pensions and National Insurance report, dated 12 October 1958, National Archives, PIN 26/21799 part 2 (P1030498).

80. Illegible signature, dated 12 November 1958, National Archives, PIN 26/21799 part 2 (P1030499).

81. René Leriche, *The Surgery of Pain* (trans. Archibald Young) (London: Ballière, Tindall and Cox, 1939), 202.

82. Leon Gillis, *Artificial Limbs* (London: Pitman Medical Publishing, 1957), 437.
83. This was also the concern of Randall, Ewalt and Blair, 'Psychiatric Reactions to Amputation', 645.
84. D.S. McKenzie, 'The Elderly Amputee', *British Medical Journal* (1953): 153–5.
85. Signature illegible, 15 August 1956, National Archives, PIN 26/21799.
86. Gillis, *Amputations*, 339.
87. Note by Leon Gillis, dated 17 October 1956, National Archives, PIN 26/21799.
88. Tristram Samuel (Medical Superintendent, Queen Mary's Hospital), 'Summary of Case', 16 November 1956, National Archives, PIN 26/21799.
89. 'Sonalgin' (advertisement), *Transactions of the Association of Industrial Medical Officers*, 7(3) (1957): vi.
90. Samuel, 'Summary of Case', 16 November 1956, National Archives, PIN 26/21799.
91. Ministry of Pensions and National Insurance, examination, signature illegible, 12 November 1958, National Archives, PIN 26/21799.
92. Letter from Dr I.H.M. Curwen to Dr Bryans, 20 March 1964, National Archives, PIN 26/21799.
93. Dr Ian Curwen's report, dated 27 May 1964, National Archives, PIN 26/21799.
94. Report from Queen Mary's Hospital, 22 July 1964, National Archives, PIN 26/21799.
95. W. Ritchie Russell presenting his research in 'Discussion on the Treatment of Intractable Pain', *Proceedings of the Royal Society of Medicine*, 52 (1959): 984–7.
96. Dr G. Caithness in assessment report, 29 January 1973, National Archives, PIN 26/21799.
97. Letter from Dr G. Caithness to Dr I. Curwen, dated 21 February 1974, National Archives, PIN 26/21799; letter from Hopkinson to the Welfare Officer, Irene House, Balham, dated 21 July 1974, National Archives, PIN 26/21799; letter from Ian H.M. Curwen to Dr G. Caithness, 10 April 1974, National Archives, PIN 26/21799; Medical Report by Dr Bell, 29 July 1974, National Archives, PIN 26/21799.
98. C.G. Stringer, of the Ministry of Social Security, 13 February 1974, National Archives, PIN 26/21799.
99. Letter from Hopkinson to Mr Baker, 15 January 1974, National Archives, PIN 26/21799. See also National Archives PIN 26/21799 part 2 (P1030840).
100. 'Medical Report on an Officer or Nurse Claiming Disability in Respect of Service in the Great War... Lieut. Hopkinson', 4 September 1932, National Archives, PIN 26/21799. He makes a similar statement about pain making him 'shout out' on 3 October 1939, National Archives, PIN 26/21799 (P1030602)
101. 'Supplementary Report to be Completed in Mental and Neurological Cases', 2 January 1934, National Archives PIN 26/21799 part 2 (P1040104-P1040106).

6
The Emergence of Chronic Pain: Phantom Limbs, Subjective Experience and Pain Management in Post-War West Germany

Wilfried Witte

Chronic pain in medicine

The complexity of chronic pain is a challenge for therapists. In order to meet its needs appropriately, any textbook of chronic pain therapy now states that monotherapy by a single medical discipline should not be used. Instead, so-called multimodal pain therapy, which should be interdisciplinary, is demanded today. However, this aim only reached the medical public to a greater degree by the end of the 1970s. The ways in which pain was handled were subjected to an historical transformation.[1] Over the centuries, the focus has not always been on defeating pain or even on completely removing it. The significance of chronic pain was also characterised approvingly as meaningful at times.[2]

Chronic pain has been managed in various ways, both by patients and by therapists. In natural science-oriented medicine of the nineteenth century, it was particularly difficult to gain an idea of chronic pain when no (patho-)physiological explanation appeared to be possible. But pain without lesion was not an unknown phenomenon in Western medicine, at least by the beginning of the nineteenth century.[3] The reductionistic neuroscientific idea that objectifiable *physical pain* could be separated from subjective *mental pain* has been fundamentally criticised.[4] This idea dates to the late nineteenth century without entirely representing the therapeutic orientation to pain at that time in its entirety.[5] The American occupational medicine and military physician William Livingston no longer wanted to accept the dualistic differentiation as early as 1943: 'To classify certain types of pain as "psychic" pain is purely arbitrary, because all pain is a psychic perception.'[6]

Starting in the 1960s, the idea that pain should be regarded not only as a *sensation* but also as an *emotion* gradually gained acceptance.[7]

Only with this prerequisite did it become possible to understand chronic pain not merely as a symptom but – in the case of chronification – as an independent illness. This was already formulated decades earlier in France as a demand for medicine. At the beginning of the twentieth century, the French surgeon René Leriche promoted the importance of the sympathetic nervous system for people in pain. He was the first to assume a 'pain disease' ('douleur-maladie') in cases of long-lasting pain. According to him:

> For doctors who live in constant contact with the sick, pain is simply an incident, irksome, troublesome, and frequently difficult to suppress symptom ... The number of illnesses which pain reveals is rare, and it frequently only serves to confuse us about the illness it accompanies. In contrast, in the case of a few chronic conditions, pain appears to be the entire illness, without which the problem would not exist.[8]

At the same time, the German neurologist and neurosurgeon Otfrid Foerster from Breslau, who also explored pain in the sympathetic system, emphasised that pain should be considered a 'mental experience' ('psychisches Erlebnis'): 'Pain belongs to that group of psychic phenomena that are generally characterised by feelings or affects, as opposed to sensations and perceptions.' For this reason, the proper term is 'feeling of pain' ('Schmerzgefühl') rather than 'sensation of pain' ('Schmerzempfindung'). Foerster considered the stimulus propagation in the (afferent) nervous system to be the 'physical correlate' of the feeling of pain.[9] In this way, pain itself was still somatic, physiological and objective, and could in principle be measured. This view was in contrast to Leriche's idea of 'douleur-maladie'.[10] In German pain research and medicine after the First World War, bodily experienced pain could be seen as a psychic experience.[11] But the persisting opinion was that it was a symptom and was physiological. In the case of phantom pain, this conviction was problematic from the beginning.

Pain in Nazi Germany and in post-war West Germany

National Socialism did not care much about pain. In the National Socialist period, it was inevitable that the brutality and dehumanisation of the regime would at least be accompanied by the demand of staying in control of oneself when faced with pain. This also corresponded to the body image of the Third Reich.[12] In 1935/6, a German internist and medical historian stated that physicians 'were familiar with some states of the human organism' being 'without a definite anatomical or chemical basis', thereby sentencing 'those who suffer from it to complete inactivity'. However, everything could be 'returned to a state of balance' by 'one's own mental efforts of self-education and overcoming oneself'. Unfortunately, the 'hysterical pain sensitivity' of both genders, the historian stated, had 'increased similarly to an avalanche'.[13] On the other hand, the writings of the surgeon Ferdinand Sauerbruch (1875–1951) and the educationalist Hans Wenke (1903–71) from 1936 were much more differentiated. Wenke in particular was able to explore pain and the experience of pain in a finely tuned manner, for instance, when he wrote about pain without lesion:

> At the same time, however, there is non-localised physical pain as an expression of an overall state in which it seemingly extends throughout all vital experiences. Any life activity is felt to be painful, for instance when the sufferer complains: 'Everything hurts, I feel battered, etc.'[14]

The document was re-published in post-war Germany, revised by Wenke to add recent physiological results and to eliminate all of his Nazi convictions. It may have been suitable for providing an idea of pain events. However, therapy was still oriented towards acute pain.[15]

This was all the more true in the Second World War. All options were utilised to manage acute situations – pharmacotherapy, volatile and intravenous anaesthesia, as well as the application of local anaesthetics. There was a tendency to concentrate on technical aspects, for instance, the surgical procedures best suited for amputations.[16] After the war, 'follow-up surgeries' were more common following amputations. As a consequence of the war, specific pain disorders were a central focus of the experts in the 1950s: causalgia – a special form of chronic burning pain due to nerve injuries (today known as complex regional pain syndrome type II) – and phantom pain following amputations or amputation injuries.[17]

The situation of patients who suffered from phantom pain after amputations was difficult after the war. The estimated number of unreported cases of suicide was high.[18] It was attested to individual phantom-pain patients that they were faced with pain that no longer had 'the significance of a symptom of illness and a warning that served a purpose', but 'had become the illness itself'.[19] However, therapy attempts were often destructive (neurosurgically or with alcohol instillation) and nonetheless unsuccessful. The underlying idea that the transmission of pain could be stopped by using neurosurgery was confirmed only in rare cases.

The spreading culturally pessimistic attitude that the 'pain sensitivity' of 'modern humans' had increased considerably also led to corresponding interpretations of the situation of phantom pain patients. At the end of the 1950s, for instance, it was said that those of Napoleon's soldiers whose arms or legs were amputated by military physician Dominique Larrey had sometimes borne it stoically. In modern phantom pain patients, on the other hand, there was said to be 'abnormal affective processing of the phantom experience', since they were unable to 'get over' the loss of the limbs.[20] Psychiatric therapies were attempted in this period.[21]

Amputations and their consequences had quickly become a theme in medicine in the period after the war. However, the primary focus was on so-called immediate prosthetic care (immediate rehabilitation after amputation). This corresponded to the actual or assumed desire of the amputees to be fully able to work, despite their stumps, so that they could present themselves as 'normal members' of society.[22] Focusing on restoring the ability to work followed the tradition of 'cripple care' since the First World War in Germany.[23] After the Second World War, Michel Berlemont from Berck-Plage, France was the first to publicise 'immediate care', followed by Marian A. Weiss from Warsaw, who presented the concept at the International Orthopaedists' Conference in Copenhagen in 1963. The techniques were spread via London and the USA, and also reached West Germany.[24]

Technical and orthopaedic care, as well as walking training, played a major role in the question of whether leg amputees would have to fight pain over a long period.[25] Orthopaedic care locations ('Orthopädische Versorgungsstellen') had already been set up in Germany after the First World War.[26] In (West) Berlin, there was a facility providing walking training at the Orthopaedic Care location of the Senate from 1952. It was shut down in 1968 because prosthetic care of 'amputated war victims' was deemed sufficient.[27] The number of amputees was still high at that time:

Total number of war injured persons requiring orthopaedic care	449,813
Number of war injured persons with injuries from the First World War	55,235
Persons with one leg amputated	115,489

Persons with one arm amputated	36,888
Persons with both legs amputated	9,465
Persons with both arms amputated	896
Other double amputees	1,089
Triple amputees	96
Quadruple amputees	18
Total number of amputees	163,941[28]

The medical debate about phantom pain, 1968–70

The majority of German citizens had participated enthusiastically in the war, or at least joined the ranks without complaint. In the 'economic miracle' of Germany after the war, the willingness to deal with its own past was low. The assumedly 'whole' world was not to be disturbed.[29] War and destruction were not supposed to have left lasting psychological consequences behind in the survivors. Psychodynamic theories coursing in expert circles in the Federal Republic stated that there were no limits to the 'human capacity to process violent events'.[30] The knowledge that the 'national socialistic policy of persecution and destruction' had traumatic consequences for its victims only spread gradually in the Federal Republic in the 1970s.[31] As the 'public memory of the violence of the national socialist war of destruction' slowly began to change, earlier in the 1960s, the 'public rules concerning what could be said' by former Wehrmacht soldiers, who had previously remained largely silent about the war, also changed.[32] Victim and perpetrator attributions arose once it was acceptable to talk publicly about what had occurred in the period up to 1945. 'The emotional overload that was typical of the post-war society of the federal republic, and that had occurred because of suffering simultaneously caused and experienced' was only gradually perceived.[33] In this time of emerging cultural change, the face of chronic pain also altered. Phantom pain was regarded as exceptional in this process. The words of a medical expert clarified this in April 1969: 'There is no other eerie pain syndrome whose causes are similarly unclear in medicine' (R. Dederich).[34]

At the end of the 1960s, the responsible federal labour ministry finally put the subject of phantom and stump pain on the agenda. Christian Democratic minister Hans Katzer (Minister of Labour, 1965–9) decided that a committee of medical experts should study the problems of war-injured amputees. Under these auspices, he created a project that his Social Democratic successor in office, Walter Arendt (Minister of Labour 1969–76), perpetuated. Starting with the Medical Experts Committee, which resided in the Federal Ministry of Labour, a 'Sub-committee for Stump Nerve Pain of Amputees' was created.

On 27 November 1968, this Sub-committee first met in Bonn at the Federal Ministry for Labour and Social Affairs. It included members of the medical advisory committee of the ministry as well as leading specialists ('Fachärzte') of the universities from the fields of orthopaedics, surgery, neurology, neurosurgery, psychiatry, internal medicine, balneology and physiotherapy (mostly professors).[35] The Sub-committee had the task of 'developing a scientific compilation of the aetiology, pathogenesis and therapy of stump nerve pain'. Representatives of patients or patients themselves were not invited to the sessions. However, the Sub-committee was in constant communication with the lobby of war-injured persons – the 'Association of persons injured in war, survivors of persons who died in war, and social pensioners of Germany' ('Verband der Kriegsbeschädigten, Kriegshinterbliebenen und Sozialrentner Deutschlands' (VdK)), which was also based in Bonn. The Sub-committee met for two years.

Among the various complaints that occurred after amputations, phantom pain was a particular headache for the experts. It was difficult to explain. Its therapy was problematic. Unlike neuromas, for instance, which led to stump pain, simple curative surgical therapy was not possible. A senior neurosurgeon in Cologne stated for the record in November 1969 that there was 'No uniform opinion from neurosurgeons'. In therapeutic terms, another expert representative did not even regard neurosurgery as responsible. It was 'Not a field of neurosurgical therapy'.

In March 1969, a psychiatrist from Kiel lectured that psychiatric literature on the subject was 'surprisingly sparse'. He also differentiated stump pain that was localised in the amputation stump from the phantom sensation – the 'subjective experience of the extremity that was no longer present ("phantom limb")' – and this, in turn, from phantom pain, which was localised in the 'phantom limb'. Reasoning from the 1950s was revitalised. A phenomenological-psychiatric monograph from 1952, which referred to the mysterious phenomenon as 'a basic psychosomatic phenomenon that is unique to humans', was quoted. A psychogenic origin of the pain was assumed, in the sense of 'not being able to manage' the loss of the limb. A report in which 25 stories of patients who suffered from phantom pain and who had been admitted to, and examined in, a neurological clinic was referenced: '75 surgeries had been performed on these patients; without exception, they had temporary successes at best; often there was no success, or even worsening.'

The expert representative from Kiel was surprised that the psychoanalysis of the problem at hand 'had barely been initiated', 'most likely because patients are not seen by analysts'. Sigmund Freud, however, had not dealt with it at all either, he added. The sobering conclusion was that psychiatry was unable to provide a generally valid judgement. Another psychiatrist at least suggested psychotherapeutic measures such as hypnosis and autogenic training, as well as physical measures, for therapy.

In January 1970, a neurologist from Göttingen objected to the thesis that phantom pain was generally psychogenic. Instead, it ought to have been regarded as 'the result of spontaneous activity of certain areas of the central nervous system'. Psychopharmaceuticals, suggestive therapy and autogenic training were to be considered for treatment. Brain surgery was regarded as a *last resort*: 'In intractable long term phantom pain, stereotactic interventions might be performed after exhausting all other therapeutic options.'

The most important advocates of neurosurgery for pain in the period between the wars were also quoted to categorise phantom pain: René Leriche (1879–1955) from Paris and Otfrid Foerster (1873–1941) from Breslau (Wrocław). But in a neurosurgery overview from 1969, almost nothing was left of their conceptions. It was roughly pointed out that there were authors who would entirely reject a sensory physiological interpretation of phantom pain. Instead, they 'would regard it as an abnormal mental experience reaction'. In this sense, Foerster was said to have referred to phantom pain patients as 'pain hyperpaths' and Leriche as 'persons ill from pain'. Yet even Leriche's 'pain illness' was pejoratively misinterpreted.[36] Psychotherapy was also the method of choice for the alleviation of phantom pain for a surgeon from Moers in November 1969. While surgery on the nerve strand of the sympathetic nervous system of the spine (sympathetic chain blockades and severing the sympathetic chain) had been shown to be 'useful', he was unable to make statements about interventions on the brain due to his lack of experience.[37]

Subjective experience

While medical experts sat down to document the state of knowledge, the VdK was not passive. On 18 and 19 April 1969, the second federal VdK conference on amputee matters took place in Bonn. An affected person reported his experiences, and an extract from various submissions by affected persons that had been received by the association journal *Die Fackel* (*The Torch*) was presented.[38] These letters were evidently passed on to the committee by the chief editior of *Die Fackel*. Insofar as could be discerned, however, they were not included in the considerations of the committee. The debate ran in parallel and not in a communicative exchange.

The chief editor of *Die Fackel*, Hanns Anders, had collected submissions for the conference in April 1969. In December 1967, a former soldier wrote:

> I'm amputated at the thigh and have always had severe pain, which has now become unbearable. I have had two further amputations for this, as well as surgery on the sciatic nerve. Clinical treatment likewise did not bring relief. Can a comrade who also has this pain give me the name of a doctor or hospital where he was freed from this pain?

A medical councillor (*Medizinalrat*), who was not in service, prescribed himself fractioned 'homeopathic morphine doses' without major success. At the same time, he warned against the danger of morphinism. An orthopaedist who had himself been affected for 25 years traced his pain attacks to weather sensitivity, which he sought to stop by sitting in a self-built Faraday cage. Another amputee complained about the helplessness of general practitioners: 'They prescribe medications, and when there are too many of them, they point out to the patient that the health authorities will cause difficulties when larger quantities are exceeded.' Other measures also tended to be useless; none was able to bring relief:

> The doctors of the care agencies prescribe treatments that do improve general well-being, but do not specifically address the problem in itself. In the age of heart transplants, can there really be no true help for those who were injured in war and have suffered from severe pain for decades (I have about 3 to 4 pain-free days a month)?

The VdK did not feel that the Sub-committee, whose work it welcomed in principle, was working quickly enough. Many thousands of amputees were still leaving their working lives due to the 'partly horrible pain', since they were no longer able to engage fully in their work. A more precise record of amputees suffering from chronic pain was needed in order to gain an idea of the situation. It recommended that specialised departments 'for treating severe cases' be created at the care institutions and care hospitals. The VdK quoted the wife of an affected man who had written: 'It takes a lot to still say yes to such a life at all.'

In addition, amputees suffered from a general lack of sympathy and understanding. One affected person stated: 'The worst thing was that various superiors and even doctors thought I was faking.'[39] Medical histories were complex and often a record of helplessness, since it was not possible to find a therapist who knew how to help in difficult cases. For example, a patient from Stuttgart-Untertürkheim wrote in August 1970:

> I won't describe the many curative treatments with baths, trying out new products and medications, and psychotherapeutic treatments in further detail. At the beginning of the muscular atrophy in the 'Gritti sock' and the resultant constriction of the capillaries, we had reasonably good successes with oxygen insufflations every summer (200–220 ccm into the artery three times a week), but even up to 400 ccm no longer helped after further constriction! I was only able to make it through repeated attempts to widen the arteries again by increasing the blood pressure to a certain pressure level. – When shortening the stump further in Bonn likewise did not promise lasting improvements, I was willing to declare my consent to deactivating the pain centre in the thalamus as a last resort. After HE. [unidentified abbreviation] Prof. Dr. Krayenbühl at the Zurich Canton Hospital told me about the consequences and 'that one case had already been successful', I decided not to pursue this ... The illustrated magazine 'Bunte' had a treatise on successes using procaine [local anaesthetic] injections. The pension office approved this for me, but it did not have the slightest effect.[40]

The VdK functioned as a relay station for accounts of affected persons, where they came up, taking this information to the Ministry and its Sub-committee. While the Sub-committee explored the few known options for patient care in an assumedly factual atmosphere, the VdK often received unfiltered information about the lives and feelings of those whom it represented. The desperation of being faced with individually unbearable life situations sometimes changed to anger. In a letter of the VdK to the Federal Ministry of Labour in March 1970, an amputee lamented: 'It's too bad that no one in this committee has this pain himself – surely we would be further ahead by now in that case.'[41]

The final report of 1970

The final recommendations of the Sub-committee of the Federal Ministry of Labour were published in writing in November 1970 in the official organ of German physicians, the *Deutsches Ärzteblatt* (*German Physicians' Gazette*).[42] At this time, a research mandate had also been given to the Orthopaedic Hospital of the University of Cologne 'with the objective of exploring further and new treatment options for amputees suffering from severe stump pain'.

As a reaction to the publication in the *Deutsches Ärzteblatt*, an anaesthetist who was the chief physician of the anaesthesia department at the Regional Hospital of Starnberg near Munich, Ottheinz Schulte-Steinberg, spoke up in November. Schulte-Steinberg complained that in comparison to the composition of the 'currently leading committee for pain problems, the Pain Clinic of Washington State University in Seattle (Prof. Bonica)', the Sub-committee was not complete without representation from an experienced anaesthetist. In a Pain Clinic, the patient would be 'jointly examined by the group of various specialists and a therapy plan would be drafted'. Blockades of the sympathetic nervous system, other nerve blockades and differentiated spinal anaesthesia could often provide help. Surgeries would not be needed at all in this case.[43]

In April 1971, Schulte-Steinberg received a response from the Ministry. The responsible 'Medizinalrat' replied that he wanted to provide the Cologne 'research team' with his 'very interesting' recommendations. The Sub-committee of the Medical Expert Advisory Committee at the Federal Ministry of Labour had, however, already ceased its work: 'Further specialists can therefore no longer be consulted.'[44]

The medical experts who had been appointed to the Sub-committee were all established specialists in leading positions in university medicine who worked in established and inflexibly specialised fields. At that time, they were the group of physicians who enjoyed extraordinary prestige (they were dubbed 'demi-gods') and they had the highest income.[45] Anaesthetists were not among them. The negotiations of these experts between 1968 and 1970 show the image of a definite distance from the worries and suffering of the patients whose illness they were debating. Their social situation was not brought up; the psychological situation remained suspect. Allegedly psychotic constellations were held responsible for the development of the illness in terms of differential diagnostics. Patients individually tried to be heard through their association, but in the end felt abandoned by the medical and scientific world.

In the bulletin that it had eventually formulated, the Sub-committee itself had noted that one doctor alone seldom had a chance of helping a stump pain patient in an individual case. This was applied both to specialists and to general practitioners: 'In many cases, only consistent and systematic treatment in which physicians from several fields must cooperate can correct or decisively relieve this pain.'[46] The Minister of Labour added that instead of the 'side-by-side' arrangement of the VdK and medical expertise, a search for solutions should occur 'on the widest possible basis'.[47]

Hopes were focused on the 'research project systematically to investigate amputees who suffer from stump and phantom pain', which was financed by the federal government. Patients were intended to have the opportunity to make statements therein. The project was initially delayed and did not commence until 1972. Due to 'the lack of a data processing system', the originally selected Orthopaedic Hospital of the University of Cologne had deemed itself unable to implement it. The Psychiatric University Hospital of Würzburg took over the project instead. In Würzburg, a decision was made to question and examine thigh-level amputees.[48] Data collection did not start until 1973 and was completed in 1976. The final report was published in 1979.[49] A total of 142 thigh-level amputees had been examined and 583 had been questioned in writing. All of them – even the affected persons – had hoped for a breakthrough, a silver bullet, since the highest expertise, represented by university specialists, had been guaranteed. However, the results of the final report were again monotherapeutically oriented and evidently disappointing, especially for the VdK:

> Even according to the results of our investigation, nothing speaks for the assumption that a single, specific treatment method might make it possible reliably to improve or even remove the late syndrome that follows amputations, either as a whole or in part.[50]

It is doubtful whether anaesthesiology could have entirely changed the negotiations in the period 1968–1970. However, it seems that federal German policies did not even have anaesthesiology on their agenda as a medical discipline to be heard. While Schulte-Steinberg himself did address chronic pain later, he was primarily considered to be one of the few specialists in surgical regional anaesthesia in the Federal Republic.[51] Stump and phantom pain therapy initially remained anchored in the old form of disciplinary thought: 'Pain therapy was originally regarded as a field of medical work that all fields claimed for themselves.'[52] This also meant that all fields separately attempted to provide therapy for patients with pain problems. Therefore, the 'treatment of pain' was among the 'basic duties of a physician'. The 'medical gaze' (Foucault) had to be trained. Patient subjectivity took second place; the main symptom was regarded as objectifiable: 'It is all the more astonishing', a German neurologist adjudged in 1969, 'how uncertain physicians are when pain is the main symptom and the diagnostic approach is to be derived from its analysis.'[53]

The Sub-committee for Stump Nerve Pain of Amputees (1968–1970) and the Federal Ministry of Labour had explored basic ideas about turning around the treatment of patients who had phantom pain. The final statements were that a single specialist authority should no longer judge the illness and make its judgement based on symptoms. Instead, physicians from various disciplines were to interact, allowing patients to be heard and various groups of participants – including representatives of various interests in addition to the doctors – to be involved. In other words, the statements of the affected persons themselves were to be recognised as part of the actual problem, so that the categorisation as 'abnormal' and deficient had to be dropped. The change, which began years later, was, however, in practice not completed, since the monotherapeutic orientation had not yet been overcome. With this, the societal theoretical orientation towards cultural pessimism remained.

Pain management and cultural pessimism

Anaesthesiology has a special position among the medical fields that deal with pain. The founding myth of modern anaesthesia is based on the idea that pain that was previously inflicted and experienced in the surgical operating room had been lifted from humanity.[54] Pain during surgery was no longer taken for granted. The idea that pain is necessary for survival in the operating room has become obsolete.[55] Only recently, the readers of the influential *New England Journal of Medicine* voted the description of the first anaesthesia of 16 October 1846 as 'the most important article in *NEJM* history' on the event of the 200th anniversary of the journal.[56] However, this initially did not affect chronic pain. For the USA, it was also proven that the introduction of anaesthesia in the nineteenth century did not mean that everyone received surgery under anaesthesia from that time on.[57]

The practical turn in chronic pain therapy is primarily traced back to American anaesthetist John Bonica. In the early 1950s, he first tried to found the *management of pain* on the methods of regional anaesthesia.[58] In 1953, he published a book entitled *The Management of Pain*, consisting of over 1,500 pages. Focusing on regional anaesthesia, it was intended to list all known options for treating pain.[59] According to today's knowledge, regional anaesthesia alone is not the way to obtain regular therapeutic success in cases of chronic pain.[60] However, the thesis of a practical change nonetheless emphasises the significance of Bonica's almost cookbook-like compendium. After all, it started a move towards detachment from the components of cultural theories that, while suitable for wordy explorations of chronic pain, only considered the statements of patients who were suffering from chronic pain in a distanced manner, in practice leaving patients to themselves with their pain.

From the commencement of his professorship at the University of Washington in Seattle in 1961, Bonica was able to begin to realise his concept of a pain clinic with a multidisciplinary structure. The pain clinic was primarily for outpatients and was, in the beginning, a so-called *nerve block clinic*.[61] Bonica's underlying thoughts were, on the one hand, that clinical pain had to be differentiated from pain that was created in a laboratory and, on the other hand, acute pain had to be distinguished from chronic pain, which no longer exercised a warning function like pain in acute situations.[62] The decisive theoretical expansion of Bonica's approach was the 'Gate Control Theory' of British physiologist Patrick D. Wall and Canadian psychologist Ronald Melzack from 1965.[63] Even though the theory in itself is no longer significant, the metaphor of the gate through which incoming and outgoing stimuli must pass to trigger or modify a pain sensation was enormously influential.[64] It was possible to use it for theoretical depictions of social and psychological influences as being constitutive of pain sensations: 'The gate control theory of pain provided the physiologic basis for the biopsychosocial model.'[65] Bonica's concept of a pain clinic was further therapeutically expanded in 1968 by the theoretical behavioural learning approach of clinical psychologist Wilbert E. Fordyce.[66] Bonica tirelessly propagated his concept. He did this internationally. Successes began only in the 1970s.

The position of anaesthesiology was difficult in Germany. It was only able to establish itself as an independent discipline late, starting in the 1960s. Unlike British anaesthesia, for example, which came from the general practitioner's field of work, German anaesthesia originated in surgical personnel.[67] German surgeons fought for a long time before they gradually allowed an independent West German anaesthesia.[68]

When one of the first German professors of anaesthesiology, Rudolf Frey, obtained his teaching chair in Mainz in 1960, he chose pain treatment as the subject of his first lecture. This made sense, since anaesthesia was subject to the idea of having won over pain. Following customary habits, Frey complained in May 1961 that 'today's human beings' had also become 'more sensitive' when it came to bearing pain because of their 'more sophisticated life habits'. However, he immediately emphasised his core discipline of anaesthesia by pointing out that no one had to endure surgery without anaesthesia any longer.[69] On the other hand, he quickly expanded his knowledge and picked up newer developments in his field from overseas. Starting in 1970, he gave chronic pain therapy room to unfold at his clinic by founding the first German pain clinic, following the American example.[70] In originally modest conditions, an outpatient nerve blockade department was founded there; however, Frey himself did not engage in practical work in this department.[71] His task was to propagate the new approach in speech and writing. When he addressed the subject of fighting pain in a speech at which Bonica was present in 1974, he regarded anaesthesiology as having the duty of also dealing with chronic pain: 'Out of this thousand-fold experience in removing pain to allow surgical interventions, anaesthesia has always had to deal with the treatment of chronic, and otherwise therapy-resistant states of pain as well.'[72]

However, pessimistic cultural approaches remained in vogue to compete with it. In particular, psychosomatics – which later became a central component of chronic pain therapy – made use of them. The narrative that the entirety of society should be 'anaesthetised' because modern humans were no longer able to bear suffering or did not want to do so contradicted that of anaesthesia, which wished to have the means of overcoming any pain.[73] In 1965, for instance, Hamburg internist and psychosomatic physician Arthur Jores addressed pain as a phenomenon. Jores clearly differentiated acute pain from chronic pain, which he described as notable because of its problematic qualities. In a culturally pessimistic manner, however, he also attested that anaesthesia and 'pain relieving medications' had led to a situation in which 'today's people fear pain and therefore experience it more intensively when it occurs'.[74] A few years later, in 1972, a psychiatrist categorised this development as fatal: physicians were degraded into 'technicians fighting pain'; science was merely abstracting partial aspects; and patients saw pain only as a 'nuisance disturbance of operations in the organism', which had to be corrected immediately in a similar fashion to a 'defect in an automobile'. He stated that this led to the loss of the human aspect in human pain.[75]

Even when chronic pain therapy started to become established in West Germany after the American example, it initially remained common practice to integrate the culturally pessimistic warning in parallel with wider depictions of pain as a medical subject. In a work from 1977, an otherwise entirely unknown author diagnosed the 'inability of modern human beings to bear pain' once again.[76]

Neurologist and philosopher Viktor von Weizsäcker (1886–1957) played a central role in German psychosomatics. Von Weizsäcker is regarded as the founder of the 'Heidelberg School' of psychosomatics. He founded an holistic consideration of the doctor–patient relationship in his *Medizinische Anthropologie* (*Medical Anthropology*). He made numerous statements about pain, but explicitly in 1926/1927, 1936 and 1951.[77] In later debates on pain, his 1926/1927 intervention was most influential. In that essay, he wrote that the choice of becoming a physician found its meaning in 'turning toward pain'. Pain involved a choice 'between being strong or weak against the pain' at an early time. The doctor was to be a 'pain expert', adept at differentiating between 'destructive pain' and 'constructive pain': the latter would have to remain, while the former would need to be relieved. Removal of pain, he said, was not the sense and purpose of the doctor–patient relationship: 'For both the doctor and the patient, the task is to manage the pain work and to make a decision about it.'[78] He did not express himself more concretely. Followers of his *Medizinische Anthropologie* also admitted that it contained few instructions for 'concrete action'.[79]

Even decades later, after the turn of cultural pessimism, Weizsäcker's statements remained popular. In a speech given to pain therapists and interested parties in May 1984, a prominent medical historian from Heidelberg who had studied neurology and psychiatry stated that it was true that one had to doubt whether some kinds of pain actually made sense. However, he also brought up von Weizsäcker's theory of constructive and destructive pain, and stated pessimistically:

> And while pain was always a personal challenge that had to provoke attitudes such as patience, courage, endurance, humility and resignation, it is now made into a political problem, which gives rise to an avalanche of anaesthesia consumers. It seems that pain only serves to generate a demand for more and more medications.[80]

These opinions pointed to a tradition of Christian metaphysics, whereas pain management at least did not need to be ascribed to that tradition. Within the same years in which anaesthesiology made the practical turn in (chronic) pain therapy its own, the orientation to dealing with the subject of pain according to historical ideas lost significance for the field. This occurred at a time in which German philosophy also underwent fundamental changes by opening up to Anglo-Saxon discourses. One might risk the thesis that the practical turn in pain therapy corresponded to the linguistic turn in philosophy.[81] The ways in which one spoke about pain patients changed. The narratives of pain patients gradually gained significance, while language shifted into focus in terms of the theory of cognition. Chronic pain management does not cure those who are ill, but manages them. Helplessness and passivity are rejected in this construct. Pain cannot be removed; patients have to learn to manage it. However, there are no limits to this process in principle. In 1997, though, American psychologist Robert Kugelmann recognised that the self-technologies of pain rendered social protest in itself silent:

> If pain is truly epidemic today, then something is terribly wrong, not only with patients, or 'inadequate' pain technologies, but with the social matrix that produces suffering. To tempt people in pain to be co-managers in such a social world only deepens our true helplessness.[82]

In this regard, one might ask whether chronic pain therapy is affirmative at its core. In historical terms, turning away from practical therapeutic contact with patients does not appear to be a reasonable option. Chronic pain was not again to be declared an intangible enigma, to be explored only by higher ideas that call for humility and passiveness. Still, this does not affect critical questions concerning the concept of the management of pain.

Stump and phantom pain caused by war injuries decreased greatly in recent decades as many patients passed away in the meantime. The spectrum of phantom pain patients has shifted to vascular patients who receive amputations. The 'Amputierten-Initiative e.V.' ('Amputees' Initiative e.V.'), a 'federal association for arm and leg amputees and persons with vascular disorders', which is independent of prosthesis manufacturers, was founded by affected persons Dagmar Gail and Henry Ziemendorf (d. 1995 in Berlin) in 1991.[83] The initiative sees itself as a mediation and advice location for amputees or persons threatened by amputation.[84] The relation between established chronic pain therapy and the representation of interests of phantom pain patients, however, is still affected by tension today. Yet therapeutic options have by now expanded significantly. That expansion began with the practical change in pain therapy.

Notes

1. Joanna Bourke, 'Pain and the Politics of Sympathy, Historical Reflections, 1760s to 1960s' (University of Utrecht, 2011), available at: http://dspace.library.

uu.nl/bitstream/handle/1874/210217/Bourke_Joanna_oratie.pdf?sequence=1 (date accessed 21 February 2014).

2. Esther Cohen, *The Modulated Scream: Pain in Late Medieval Culture* (University of Chicago Press, 2010); Javier Moscoso, *Pain: A Cultural History* (Basingstoke: Palgrave Macmillan, 2012).

3. Andrew Hodgkiss, *From Lesion to Metaphor: Chronic Pain in British, French and German Medical Writings, 1800–1914* (Amsterdam: Rodopi, 2000).

4. Roselyne Rey, *The History of Pain* (Cambridge, MA: Harvard University Press, 1995); David B. Morris, *The Culture of Pain* (Berkeley: University of California Press, 1991).

5. Daniel Goldberg, 'Pain without Lesion: Debate among American Neurologists, 1850–1900', *19: Interdisciplinary Studies in the Long Nineteenth Century*, 15 (2012).

6. William K. Livingston, *Pain Mechanisms: A Physiologic Interpretation of Causalgia and Its Related States* (New York: Macmillan, 1943), 62–80 at 70.

7. Hodgkiss, *From Lesion to Metaphor*.

8. René Leriche, *La Chirurgie de la Douleur*, 3rd edn. (Paris: Masson & Cie, 1949 [1937]), 29.

9. Otfrid Foerster, *Die Leitungsbahnen des Schmerzgefühls und die chirurgische Behandlung der Schmerzzustände* (Berlin: Urban & Schwarzenberg, 1927), 1–2.

10. Manfred Zimmermann, 'The History of Pain Concepts and Treatment before IASP', in John D. Loeser, Ronald Dubner and Harold Merskey (eds), *The Paths of Pain 1975–2005* (Seattle: IASP Press, 2005), 1–21 at 17; Wilfried Witte, 'Schmerz und Anästhesiologie. Aspekte der Entwicklung der modernen Schmerztherapie im 20. Jahrhundert', *Anaesthesist*, 60 (2011): 555–66, at 556, 559.

11. Roland Wörz and Ronald Lendle, *Schmerz – psychiatrische Aspekte und psychotherapeutische Behandlung* (Stuttgart: Gustav Fischer, 1980), 1.

12. Paula Diehl (ed.), *Körper im Nationalsozialismus.Bilder und Praxen* (Munich: Fink Schönigh, 2006).

13. Georg Sticker, 'Zur Vorgeschichte der Schmerzbehandlung', *Der Schmerz. Deutsche Zeitschrift zur Erforschung des Schmerzes und seiner Bekämpfung*, 8 (1935/6): 55–63.

14. Ferdinand Sauerbruch and Hans Wenke, *Wesen und Bedeutung des Schmerzes* (Berlin: Junker und Dünnhaupt, 1936), 64.

15. Ferdinand Sauerbruch and Hans Wenke, *Wesen und Bedeutung des Schmerzes*, 2nd edn (Frankfurt am Main: Äthenäum, 1961).

16. H.J. Lauber, 'Die Schmerzbekämpfung im Felde', *Münchener Medizinische Wochenschrift*, 89 (1942): 345–8; 'Bericht über die Arbeitstagung der Deutschen Gesellschaft für Orthopädie', *Zentralblatt für Chirurgie*, 70 (1944): 905–8; F. Michelsson, 'Erfahrungen mit der Frühamputation bei Erfrierungen', *Zentralblatt für Chirurgie*, 70 (1944): 997–1002.

17. L.A. Reynolds and E.M. Tansey (eds), *Innovation in Pain Management. The Transcript of a Witness Seminar Held by the Wellcome Trust Centre for the History of Medicine at UCL, London, on 12 December 2002* (London: Wellcome Trust Centre, 2004), xxi, 14.

18. Interview with Roland Wörz (pain therapist and psychiatrist, Bad Schönborn, Germany), 20 June 2013.

19. Voßschulte (Munich), 'Pathogenese und Behandlung chronischer Schmerzzustände. 68. Tagung der Deutschen Gesellschaft für Chirurgie in München', *Zentralblatt für Chirurgie*, 76 (1951): 620–1.

20. H. Linke, 'Schmerzprobleme unserer Zeit', *Münchener Medizinische Wochenschrift*, 100 (1958): 104–9; H. Linke, 'Schmerz und Persönlichkeit', *Die Medizinische Welt* (1959): 1692–700.

21. G.E. Störring, 'Der Schmerz aus der Sicht des Nervenarztes', *Die Medizinische Welt* (1964): 1389–97.

22. R. Dederich, 'Prothetische Sofortversorgung. Kardinalproblem ist der Schmerz des Amputationsstumpfes', in H. Anders (ed.), *Stumpfschmerzen – Geißel der Amputierten* (Bonn: Verband der Kriegsbeschädigten, Kriegshinterbliebenen und Sozialrentner Deutschlands, 1971), 23–8 at 24.

23. Philipp Osten, *Die Modellanstalt. Über den Aufbau einer 'modernen Krüppelfürsorge' 1905–1933* (Frankfurt am Main: Mabuse Verlag, 2012); Philipp Osten, 'Die Rehabilitation verwundeter Soldaten im Ersten Weltkrieg', in Jochen Henning and Udo Andraschke (eds), *Weltwissen. 300 Jahre Wissenschaften in Berlin* (Munich: Hirmer, 2010), 162.

24. Michel Berlemont, 'Ten Years of Experience with the Immediate Application of Prosthetic Devices to Amputees of the Lower Extremities on the Operation Table', *Prosthetics International*, 3 (1969): 8–18; 'Halt in Gips', *Der Spiegel*, 43 (1966): 173–4; Dederich, 'Prothetische Sofortversorgung', 23–8.

25. Herbert Kersten, *Gehschule für Beinamputierte. Ein Leitfaden für Beinamputierte, Fachärzte für Orthopädie, Orthopädiemechaniker, Krankengymnastinnen und Übungsleiter im Versehrtensport* (Stuttgart: Thieme, 1975); Gertrude Mensch and Wieland Kaphings, *Physiotherapie und Prothetik nach Amputation der unteren Extremität* (Berlin: Springer, 1998).

26. F. Blohmke, 'Entwicklung und Aufgaben der orthopädischen Versorgung', in Anders (ed.), *Stumpfschmerzen*, 50–6 at 54.

27. Letter of the 'Orthopädische Versorgungsstelle Berlin' to the 'Senatsverwaltung für Soziales IX B 12', 26 August 1994. Collection of Dagmar Gail, Amputierten-Initiative e.V. Berlin (May 2013).

28. Source: Federal Labour Ministry, Bonn, status of 30 September 1968; Blohmke, 'Entwicklung und Aufgaben', 50.

29. Patrick Kury, *Der überforderte Mensch. Eine Wissensgeschichte vom Stress zum Burnout* (Frankfurt am Main: Campus, 2012), 199–203.

30. Svenja Goltermann, *Die Gesellschaft der Überlebenden. Deutsche Kriegsheimkehrer und ihre Gewalterfahrungen im Zweiten Weltkrieg* (Munich: Pantheon, 2011), 426.

31. Ibid., 421.

32. Ibid., 448.

33. Kury, *Der überforderte Mensch*, 294.

34. Anders (ed.), *Stumpfschmerzen*, 15.

35. The following passages refer, unless indicated otherwise, to the file Bundesarchiv Koblenz B 149, No. 99855.

36. Leriche's book *Chirurgie de la Douleur* was available in a German translation: René Leriche, *Chirurgie des Schmerzes* (trans. E. Fenster) (Leipzig: Johann Ambrosius Barth, 1958).

37. Bundesarchiv Koblenz B 149, No. 99855.

38. Anders (ed.), *Stumpfschmerzen*, 15–65.

39. Hanns Anders, 'Amputierte berichten über ihre Erfahrungen (II)', in Anders (ed.), *Stumpfschmerzen*, 59–65 at 63.

40. Bundesarchiv Koblenz B 149, No. 99855. 'Gritti-Stokes-Amputation': a special technique for thigh amputation.

41. Bundesarchiv Koblenz B 149, No. 99855.
42. Ernst Goetz, 'Ätiologie, Pathogenese und Therapie der Stumpfschmerzen Amputierter', *Deutsches Ärzteblatt*, 47 (1970): 3511–15.
43. Cf. Ottheinz Schulte-Steinberg, 'Aufgabenbereiche des Anästhesisten außerhalb des Operations- und Kreißsaales', *Münchener Medizinische Wochenschrift*, 107 (1965): 1706–1712; Ottheinz Schulte-Steinberg, 'Leitungsblockaden in der modernen Anästhesie. Part I and II', *Anästhesiologische Praxis*, 2 (1967): 91–8 and 99–103.
44. Bundesarchiv Koblenz B 149, No. 99855.
45. Martin Dinges, 'Aufstieg und Fall des "Halbgottes in Weiß"? Gesellschaftliches Ansehen und Selbstverständnis von Ärzten (1800–2010)', *Medizin, Gesellschaft und Geschichte*, 31 (2013): 145–59.
46. 'Zur Ätiologie, Pathogenese und Therapie der Stumpfschmerzen Amputierter (Merkblatt)', in Anders (ed.), *Stumpfschmerzen*, 7–15 at 8.
47. Walter Arendt, 'Geleitwort', in Anders (ed.), *Stumpfschmerzen*, 5.
48. Hanns Anders, 'Forschungsvorhaben über Amputiertenschmerzen', *Fackel*, 26 (1972): 2.
49. *Ibid.*, 4; Hanns Anders, 'Forschungsauftrag Stumpfschmerzen', *Fackel*, 28 (1974): 7; Bundesarchiv Koblenz B 149, No. 99855; Friedrich Danke and Otto Schrappe, *Forschungsbericht Stumpschmerzen Amputierter* (Bonn: Bundesminister für Arbeit und Sozialordnung, 1979).
50. Danke and Schrappe, *Forschungsbericht*, 94; cf. Hanns Anders, 'VdK: 20 Jahre Impulse für Schmerzforschung', *Fackel*, 41 (1987): 13.
51. Ottheinz Schulte-Steinberg, 'Periphere Nervenblockaden in der Schmerzbehandlung', in Rudolf Frey and Hans Ulrich Gerbershagen (eds), *Schmerz und Schmerzbehandlung heute* (StuttgartGustav Fischer, 1977), 99–103. J. Schüttler and W. Schwarz, 'Erinnerungen und Berichte aus der Pionierzeit der Anästhesie an deutschen Krankenhäusern', in Jürgen Schüttler (ed.), *50 Jahre Deutsche Gesellschaft für Anästhesiologie und Intensivmedizin. Tradition & Innovation* (Berlin: Springer, 2002), 317–28 at 321. There was not even much updated literature about the subject. In 1971 a small textbook of 89 pages about regional pain therapy was published in West Germany (Hans Georg Auberger, *Regionale Schmerztherapie. Leitfaden für die Praxis* (Stuttgart: Thieme, 1971)).
52. Manfred Zimmermann and Hanne Seemann, *Der Schmerz – Ein vernachlässigtes Gebiet der Medizin? Defizite und Zukunftsperspektiven in der Bundesrepublik Deutschland* (Berlin: Springer, 1986), 138.
53. Rudolf Janzen, *Elemente der Neurologie* (Berlin: Springer 1969), 94.
54. Dennis Brindell Fradin, '*We Have Conquered Pain': The Discovery of Anesthesia* (New York: McElderry, 1996).
55. Stephanie J. Snow, *Operations without Pain: The Practice and Science of Anaesthesia in Victorian Britain* (Basingstoke: Palgrave Macmillan, 2006).
56. Henry Jacob Bigelow, 'Insensibility During Surgical Operations Produced by Inhalation', *Boston Medical and Surgical Journal*, 35 (1846): 309–17; Karen Buckley, 'The Most Important Article in NEJM History', http://blogs.nejm.org/now/index.php/the-most-important-article-in-nejm-history/2012 (date accessed 24 February 2014).
57. Martin S. Pernick, *A Calculus of Suffering: Pain, Professionalism, and Anesthesia in Nineteenth-Century America* (New York: Columbia University Press, 1985).

58. John J. Bonica, 'Management of Intractable Pain with Analgesic Block', *Journal of the American Medical Association*, 150 (1952): 1581–6; John J. Bonica, *The Management of Pain*. With Special Emphasis on the Use of Analgesic Block in Diagnosis, Prognosis, and Therapy (Philadelphia: Lea & Febiger, 1953).

59. Bonica, *Management of Pain*.

60. Cf. Hans Christoph Niesel and Hugo van Aken (eds), *Lokalanästhesie, Regionalanästhesie, Regionale Schmerztherapie*, 2nd edn (Stuttgart: Thieme Georg Verlag, 2003).

61. John J. Bonica, 'The Management of Pain', in Günther Thomalske, Erich Schmitt and Matthias Gross (eds), *Schmerzkonferenz: Ein Handbuch für Pathogenese, Klinik und Therapie des Schmerzes* (Stuttgart: Gustav Fischer, 1992–1998), looseleaf notebook, delivery 1–2 March 1984 (6.2), 19–36.

62. Isabelle Baszanger, *Inventing Pain Medicine: From the Laboratory to the Clinic* (New Brunswick: Rutgers University Press, 1998).

63. Ronald Melzack and Patrick D. Wall, 'Pain Mechanisms: A New Theory', *Science*, 150 (1965): 971–9; Marcia Meldrum, 'A Capsule History of Pain Management', *Journal of the American Medical Association*, 290 (2003): 2470–5.

64. Robert Kugelmann, 'The Psychology and Management of Pain: Gate Control as Theory and Symbol', *Theory & Psychology*, 7 (1997): 43–65. See also David Biro, Chapter 4, this volume.

65. Gordon Waddell, 'A New Clinical Model for the Treatment of Low-Back Pain', *Spine*, 12 (1987): 632–44 at 637.

66. Wilbert E. Fordyce, Roy S. Fowler and Barbara DeLateur, 'An Application of Behaviour Modification Technique to a Problem of Chronic Pain', *Behaviour Research and Therapy*, 6 (1968): 105–7; Wilbert E. Fordyce, *Behavioral Methods for Chronic Pain and Illness* (St Louis: Mosby, 1976).

67. Johan S. Pöll, *The Anaesthetist 1890–1960* (Rotterdam: Erasmus Publishing, 2011); Johan S. Pöll, 'History of Anaesthesia: Why Did Professional Anaesthetists Appear in Britain First?', *European Journal of Anaesthesiology*, 29 (2012): 405–8.

68. Heike Petermann, 'Angloamerikanische Einflüsse bei der Etablierung der Anästhesie in der Bundesrepublik Deutschland im Zeitraum von 1949–1960', *AINS*, 40 (2005): 133–41; Jürgen Schüttler (ed.), *55 Years German Society of Anaesthesiology and Intensive Care Medicine. Tradition and Innovation* (Berlin: Springer, 2012); Michael Goerig and Jochen Schulte am Esch, *Die Entwicklung des Narkosewesens in Deutschland von 1890–1930* (Lübeck: Steintor, 2012).

69. Rudolf Frey, 'Neue Wege der Schmerzbekämpfung', *Anaesthesist*, 11 (1962): 51–6.

70. Hugo A. Baar and Hans Ulrich Gerbershagen, *Schmerz, Schmerzkrankheit, Schmerzklinik* (Berlin: Springer, 1974); H.U. Gerbershagen, F. Magin and W. Scholl, 'Die Schmerzklinik als neuer Aufgabenbereich für den Anästhesisten' *Anästhesioloische Informationen*, 16 (1975): 41–4; H.U. Gerbershagen, R. Frey, F. Magin *et al.*, 'The Pain Clinic. An Interdisciplinary Team Approach to the Problem of Pain', *British Journal of Anaesthesia*, 47 (1975): 526–9; Hugo A. Baar, *Schmerzbehandlung in Praxis und Klinik* (Berlin: Springer, 1987).

71. Witte, 'Schmerz und Anästhesiologie', 562; Gholam Sehhati-Chafai (ed.), *Schmerzdiagnostik und Therapie, Volume 1. Festschrift für Rudolf Frey* (Bochum: Winkler, 1985).

72. Rudolf Frey, 'Die Schmerzbehandlung aus der Sicht des Anaesthesisten', in R. Frey, J.J. Bonica, H.U. Gerbershagen and D. Groß (eds), *Interdisziplinäre Schmerzbehandlung* (Berlin: Springer 1974), 11–12.

73. Years later the memoirs of the surgeon and anaesthetist Hans Killian under the title 'Fighting the pain' were very popular (Hans Killian, *Im Kampf gegen den Schmerz. Mein Abenteur mit der Narkose* (Munich: Kindler, 1979); Hans Killian, *Im Kampf gegen den Schmerz* (Munich: Moewig, 1982)). The book was about acute pain, but it suggested that pain as such was beaten. Killian had written memoirs before, under titles like 'Behind us is only God', 'Life or Death', 'As long as the heart beats' and so on.

74. Arthur Jores, 'Der Schmerz in der Sicht heutiger medizinischer Forschung', *Universitas. Zeitschrift für Wissenschaft, Kunst und Literatur*, 20 (1965): 1027–40.

75. Hans-Werner Janz, 'Geistesgeschichtliche Fragen zur Schmerzbekämpfung', in Rudolf Janzen, Wolf D. Keidel, Albert Herz and Carl Steichele (eds), *Schmerz. Grundlagen – Pharmakologie – Therapie* (Stuttgart: Thieme, 1972), 182–3.

76. G. Schmidt, 'Geistesgeschichtliche Aspekte der Schmerzbehandlung', in Frey and Gerbershagen (eds), *Schmerz und Schmerzbehandlung*, 46–93.

77. Viktor von Weizsäcker, 'Die Schmerzen', in Rainer-M.E. Jacobi (ed.), *Schmerz und Sprache. Zur Medizinischen Anthropologie Viktor von Weizsäckers* (Heidelberg: Universitätsverlag, 2012), 133–52; 'Zur Klinik der Schmerzen', in Peter Achilles, Dieter Janz, Martin Schrenk *et al.* (eds), *Viktor von Weizsäcker. Gesammelte Schriften*, vol. 3 (Frankfurt am Main: Suhrkamp, 1990), 537–48; 'Das Missliche am Schmerz', in Peter Achilles, Dieter Janz, Martin Schrenk *et al.* (eds), *Viktor von Weizsäcker. Gesammelte Schriften*, vol. 6 (Frankfurt am Main: Suhrkamp 1986), 504–10.

78. Weizsäcker, 'Die Schmerzen', 133–52.

79. Wilhelm Rimpau, 'Einführung in Leben und Werk Viktor von Weizsäckers', in Wilhelm Rimpau (ed.), *Viktor von Weizsäcker – Warum wird man krank? Ein Lesebuch* (Frankfurt am Main: Suhrkamp, 2008), 15–23 at 21.

80. Heinrich Schipperges, 'Vom Wesen des Schmerzes. Eine medizinhistorische und philosophische Betrachtung', in Thomalske, Schmitt snd Gross (eds), *Schmerzkonferenz*, delivery 3, November 1985 (1.1), 13–23.

81. Ernst Tugendhat, *Vorlesungen zur Einführung in die sprachanalytische Philosophie* (Frankfurt am Main: Suhrkamp, 1976); Herbert Schnädelbach, 'Nach 40 Jahren', *Information Philosophie*, 3–4 (2012): 8–16.

82. Kugelmann, 'Psychology and Management of Pain', 62. Cf. Bettina Hitzer, 'Die Therapeutisierung der Gefühle – eine Geschichte aus dem 20. Jahrhundert', *Der Mensch*, 42–3 (2011): 1, 2, 17–20; Sabine Maasen, Jens Elberfeld, Pascal Eitler and Maik Tändler (eds), *Das Beratene Selbst. Zur Genealogie der Therapeutisierung in den 'langen' Siebzigern* (Bielefeld: Transcript, 2011).

83. H. Heidrich, '15 Jahre Amputierten-Initiative e.V. Vasa', *European Journal of Vascular Medicine*, 38 (2009), Suppl. S/74: 5–8.

84. www.amputierten-initiative.de (date accessed 4 February 2014).

7
A Quantity of Suffering: Measuring Pain as Emotion in the Mid-Twentieth-Century USA

Noémi Tousignant

Rate your pain from 1 to 10. Does it matter that your 6 is not the same as another's? It does not: this number is *your* pain. It contains your memories, hopes, fears ... the unique, variable experience of your thinking-feeling self. Are you part of an experiment to compare the relief afforded by drug A with drug B? Off your pain will go, carrying in its number the imprint of your memories, hopes, fears ... to join others' pain as modulated by their unique subjectivities. Collected in sufficient numbers and under the right conditions, the numbers are aggregated as a collective barometer of relief from which differences between your pain and another's, but not your own judgement and experience, are flattened out. In the early 2000s, pain scales – simple tools using numbers, words or images to quantify experiences of pain – have become commonplace in clinical trials, nurses' toolkits and specialised health journals, at least in some parts of the world and especially in the USA. Their pervasiveness can be said to authorise pain as an 'inherently' subjective experience, permeated by affect and calculable only by the sufferer herself, in processes of allocating and evaluating treatment and care.

The authorisation of a more-than-sensory pain in health research and care has multiple historical roots. Techniques to diagnose, treat as well as quantify subjective pain have been designed, enacted, institutionalised and given meaning by diverse groups of actors. These range from health-care professionals dedicated to the management of stubborn, prolonged experiences of pain to professional and patient associations leading impassioned campaigns against insufficient and inequitable distributions of available relief. In this chapter, I take up one strand of this history by tracing quests for reliable pain-measuring technologies since the turn of the twentieth century. In particular, I examine how pain as an irreducibly subjective, and thus inevitably emotional, experience became what these technologies measured in the mid-twentieth-century USA.

I use the term 'authorisation' deliberately. It acknowledges the existence of more-than-sensory pain before, beyond and despite the use of technologies meant to measure it as such. But it also gives these technologies (along with standards of care, diagnostic categories and treatment protocols) an active historical and ontological role in tying pain to selves and thus to emotional experience. Pain measurement is an intriguing field in that its major innovations are often portrayed *not* in terms of growing technological sophistication (despite efforts to adapt techniques such as brain imagining to this purpose), but instead as the simple recognition or understanding, enshrined in low-tech tools such as numerical rating scales, that pain is a (subjective) 'experience which can be reported only by the sufferer'.[1] This entails a change in agreed perceptions of what counts as accurate and reliable measurement. Yet this change, I will show, results not only from moral and epistemological authorisation of pain as subjective (knowledge that pain is modulated by subjectivity and the right to have one's subjective reports of a subjective experience recognised as the true dimension of pain), but also requires technical and institutional authorisation, that is, the stabilisation and validation of working tools and practices for measuring subjective pain.

From the invention of the first algometers in the last decades of the nineteenth century, various instruments and procedures were meant to improve observations of pain by controlling its subjectivity. But what this subjectivity was made up of (imprecise judgement, emotionality, 'the personal') changed, as did ideas about how, at what level and with what techniques it should be controlled. Rationales for the use of pain-measuring technologies provide a window onto the historicity of objectivity.[2] But technologies are also practices, with implications not only for how pain is thought of as subjective, but also how this subjectivity is acted on/with as an experimental variable. If excluding the variable of 'emotion' was a central goal of pain measurement into the 1950s (and beyond), from the 1940s, some pain-measuring practices sought to include emotional modulation, experience and even judgement within the field of experimentation, shifting the locus of experimental control to other levels. This chapter explores why, and especially how, this happened.

Algometry: universal laws of sensation and sensory hierarchies

Algometers, despite their name, do not measure pain. They standardise and quantify the intensity of a pain-producing stimulus. Researchers can then determine the point at which a person just begins to feel pain: their threshold of pain sensation, sometimes defined as sensitivity. A quantified sensitivity is a comparable sensitivity. From the late nineteenth century, algometers were developed and used to map sensitivity on body surfaces, to study biosocial difference and to measure analgesic effect.

For psychophysicists, the measurement of pain responses was part of the study of sensation and perception. Algometers joined a growing number of meters for the senses (of smell, hearing and so on) considered useful for elucidating basic, universal mechanisms of sensation. Gustav Fechner systematised psychophysiology on the principle that sensory judgement could be correlated with the magnitude of a stimulus such as a ray of light of a particular brightness. Gail Hornstein argues that quantification was central in carving a space for experimental psychology that was methodologically distinct from other sciences of the mind, such as philosophy and phrenology. Sensation was one of the research topics that seemed apt for establishing psychology as a quantitative and experimental science.[3]

At the turn of the twentieth century, algometric data entered a heated debate between proponents of two theories of pain mechanisms. Was pain a specific sensory modality, served by its own neural apparatus, as suggested by specificity theory? Or did it require an intensive (also called summation or pattern) theory of pain as an extreme attribute of other sensations? To test these positions, psychologists and physiologists sought out techniques to apply discrete stimuli to minuscule areas of skin. In the 1880s, Blix and Goldscheider in Sweden began mapping out the distribution of sensory spots using needles, faradic current from a single electrode, narrow jets of hot and cold water, cork points and small drops of ether.[4] Max von Frey, a German physiologist, invented an instrument consisting of a wooden stick to which a hair – obtained from a man, woman, horse or hog – was attached with sealing wax and pressed to the skin until it bent to create highly precise and consistent stimuli.[5] Von Frey hairs are commonly cited by twentieth-century pain researchers as the first precision pain-measuring instrument.[6] To settle the nature of pain, which he considered 'one of the fundamental problems in sensory physiology and psychology', the American psychologist Karl Dallenbach continued the pursuit of imaginative means of inflicting precise amounts of painful stimulation.[7]

Dallenbach based the reliability of his evidence for the specificity theory of pain not only on the design of his algometer, but also on the quality of his observers. In the early psychology laboratory, it was common to use a small number of graduate students, colleagues or even the investigators themselves as observers. They were familiar with the principles of experimentation, often highly trained, and could be trusted to maintain a 'scientific attitude' of both attentiveness and detachment towards their sensations. The expertise of observers could be ascertained by the readers of published reports, in which their names, experience and individual results were often provided.[8] To settle fundamental theoretical issues of psychophysiology, results from a few good observers, who were considered as 'universal minds', could be generalised into laws of normal human pain sensation.[9]

Like other psychophysical instruments, the use of algometers was extended from the study of universal laws of sensation to the study of variations in sensitivity in the general population. This added pain sensitivity to brain size and facial angle as a measurable quantity for psychological, physical and criminal anthropologists to classify humans on scales of evolution and deviance.[10] The famous Cambridge Anthropological Expedition, an ambitious scientific mission to study the Torres Strait Islanders in 1898 in their as yet unspoilt state of 'primitiveness', included a Cattell Algometer.[11] The language and customs, as well as the 'acuity of the senses', of the Murray Islanders and Sea Dayaks were scrutinised. Responsible for the 'cutaneous senses', William McDougall obtained algometric readings proving that Murray Islanders were only half as sensitive to pain as English men and boys.[12]

Preoccupied with characterising 'knowable, measurable and predictable' criminal types, the Italian criminal anthropologists Cesare Lombroso and Salvatore Ottolenghi used algometric methods to detect reduced or absent sensibility to pain as a physiological marker of inherent criminality. Considering moral and physical insensibility to be intimately connected, Lombroso used an adapted common electric apparatus to deliver precisely quantified electric shocks.[13] His interests extended beyond criminality itself to the evolutionary ranking of human types. In the 1880s, he turned to the study of criminal women, but also made claims about the degenerate characteristics of all women. Pointing to data obtained with his algometer, he dismissed the widely held belief that women were more sensitive than men.[14] Taking up this work, Ottolenghi, whose book *Sensitivity of Women* was published in 1896, argued that 'true sensitivity' as detected by algometry could be distinguished from the superficial emotionality of women.[15]

Lombroso's ideas were taken up in the USA by Arthur MacDonald, who in 1902 and 1908 came before Senate Judiciary Committees to argue the rationale for establishing a 'psycho-physical laboratory' for the study of 'the criminal, pauper and defective classes' under the federal government.[16] The algometer was especially important in MacDonald's collection of instruments of precision, for he agreed with Lombroso on the association of physical with moral insensibility.[17] Indeed, he had designed his own instrument. It consisted of a scale that indicated pressure in grams, a rod and a disk, covered in flannel to avoid eliciting coldness, applied to the subject's temple. MacDonald presented data before the Senate that he had obtained from adults and thousands of schoolchildren, comparing sensitivity by social class, sex, nationality, season of birth, hair colour and mental ability, among other variables. His project was never materialised, but received extensive support among scientists and politicians until he was, claims the historian of education James Gilbert, dismissed as a 'crank' in the 1930s.[18]

Attempts to align differential sensitivity to pain with categories such as sex, criminality, intelligence and primitivism made social and scientific sense within a conceptual framework that rooted mental and moral abilities in measurable biological characteristics. Pain sensibility, in particular, had, since at least the eighteenth century, been associated in Anglo-American society with other kinds of attributes, including civilisation, intelligence and sympathy, used to distinguish classes of humans from each other and from animals.[19] Algometric practitioners sought to move beyond what they saw as superficial differences in subjects' pain responses to detect a stable, inner 'true' physiological sensitivity. This purified pain could resolve debates about the sensory nature of pain (and thus legitimise psychology as a science) or measure criminal susceptibility (and thus reform the criminal justice system). In the mid-1910s, algometry was also used to quantify the effects of analgesic drugs, but only two studies were published at the time. It was only from the mid-1930s onwards that significant investments were made in the measurement of pain for analgesic evaluation.

Differences in sensitivity to pain among types of people were expected by late nineteenth- and early twentieth-century measurers. In 1940, a group of researchers from Cornell University argued that these differences were an artefact of the contamination of the sensory pain threshold by elements of subjects' *reaction* to pain, influenced by mood, attention, disposition, etc. They claimed to have perfected the isolation of sensation. By delivering a strong beam of light in exposures of exactly three seconds through a fixed aperture onto a small area of a person's forehead that had been blackened with China ink, the set-up produced an easily identified result: 'heat finally "swelling" to a distinct, sharp stab of pain at the end'.[20] Initial results were astonishingly stable across subjects and trials. The creators of the 'Hardy-Wolff-Goodell dolorimeter', or radiant-pain technique, claimed this as evidence of the constancy of physiological sensitivity in all neurologically normal individuals, independent of gender, personality, fatigue or emotional state. 'Brothers under the skin', proclaimed a *Newsweek* headline.[21] The dolorimeter's promise of precision was quickly taken up in two fields: psychosomatic research and analgesic evaluation. Both increasingly well-funded since the 1930s and expanding in the 1940s and 1950s, these fields created unprecedented conditions for stabilising the practices and meanings of pain measurement. Interestingly, in neither field did this stabilisation fix the idea of a pure sensory threshold, stripped of emotional modulation or qualities.

Psychosomatics: calculating psychological modulation

For Harold G. Wolff, one of the creators of the dolorimeter and celebrated as a 'father' of psychosomatic medicine, linking pain to emotional events and predispositions required quantitative and objective – in the sense of bypassing subjects' emotional and vague expressions of pain – information about pain 'as a discrete sensation'.[22] Other psychosomatic researchers were interested in elucidating the psychological modulation of pain, but not through the measurement of a 'depersonalised' sensation. They instead looked to the dolorimeter as a tool for generating regular and intelligible variations in responses to pain that could be tied to emotional differences in types of persons. They did this in two ways. First, they modified the original Hardy-Wolff-Goodell technique. Uninterested in the sensory threshold, they defined endpoints that would represent a psychologically labile reaction to pain. From the early 1940s, William Chapman, a research fellow at Harvard Medical School, used the dolorimeter to measure the point at which subjects winced or withdrew from the heat beam.[23] In the later 1940s, Robert Malmo and his colleagues at the Allan Memorial Institute of Psychiatry measured autonomic responses to painful stimulation (such as sweating, finger tremor, increased pulse and so on).[24] Finally, in the 1950s, the psychologist Asenath Petrie, then at Harvard Medical School, measured the point at which a subject asked for the painful stimulation to stop, understood as the limit of their pain endurance.[25]

The results varied, but what did that mean? Psychosomatic researchers' second innovation was to pair up the dolorimeter with available personality-defining techniques and methods for measuring the physiological correlates of emotions. The availability of these techniques, from lobotomised subjects to personality assessment tools, and of agreed meanings about the psychological modulation they represented mediated the influence of the broader emergence of American psychosomatic research on conceptions of the kind of pain that could and should be measured.

Psychosomatic research and medicine was increasingly well defined, organised and funded as a field in the USA from the 1930s.[26] These developments were tied to interests in elucidating two issues: the psychological basis of differential susceptibility to chronic illness and the somatic effects of emotions and mental illness.[27] Researchers took up pain-measuring techniques such as the dolorimeter to ask related questions: how do alterations, pathologies and differences in personality affect individual responses to pain? Many of their studies can be linked through funding, publications and conferences to institutions of psychosomatic research. But perhaps more consequential was how investments in psychosomatic research generated meaningful, manipulable and measurable psychological states and variables that could be tied to quantifiable variations in reactions to pain.

By the 1940s, 'psychoneurotic' and 'lobotomised' persons were available as subjects with defined personality characteristics. These categories of person were created by psychiatric diagnostic and psychosurgical techniques. Personality tests such as the Minnesota Multiphasic Personality Inventory (MMPI) and items of the Maudsley Personality Inventory were developed from the late 1930s and were expanded in the 1940s. They were used to define lobotomy procedures in terms of their personality-altering effects and to define individuals diagnosed with psychoneuroses as disordered personalities.[28] Lobotomy was also associated with the psychological modulation of pain. Patients operated for intractable pain from the early 1940s were observed to still perceive pain, but no longer to suffer from it.[29] Measurements of pain thresholds before and after surgery showing altered reactions to, but stable perceptions of, pain were taken as illustrative of the dissociation between the sensory and emotional components of pain, as well as the personality-dependent nature of the pain-reaction threshold.[30]

Before measuring the pain reactions of 'patients following frontal lobotomy', Chapman had enlisted 'psychoneurotic patients' to clarify the variation in pain thresholds he had obtained among 'normal' subjects.[31] The difference between the groups confirmed the influence of psychological factors on reactivity to pain. Chapman's work was seen as useful for distinguishing between the physiological and psychological bases of pathologies. He was recruited to an investigation into the potential psychosomatic basis of neurocirculatory asthenia (NCA) contracted by the Office of Scientific Research and Development (a federal agency created to coordinate wartime research) to a group led by Paul D. White and the well-known psychosomatic researcher Stanley Cobb.[32] Diagnosed in many soldiers and recruits, NCA's etiology was contested and of military interest. Pain measurements helped demonstrate the influence of psychological factors in this condition.

Similarly, Malmo's team took up the dolorimeter to compare psychoneurotic and normal responses to stress. They saw standardised pain stimulation as a means of creating a standard stress situation. Levels of emotional and physical stress could be quantified through physiological correlates – lymphocyte counts, finger tremor, galvanic skin resistance and the electroencephalography (EEG) – and thus enabling a more exact measurement of the mind-body interactions that characterised psychoneurosis.[33] Their techniques for measuring somatic expressions of emotional states can be traced to research on the physiology of emotions.[34]

Asenath Petrie used personality tests directly to sort her subjects. Observing changed attitudes towards pain after lobotomies, she reports, sparked her interest in correlating personality and pain in normal subjects.[35] Her understanding of personality and of how to break it down into measurable components was based on the work of the psychologist Hans J. Eysenck, which began during the Second World War in Britain.[36] Petrie identified one of Eysenck's tests – which by the 1950s had been systematised as the Maudsley Personality Inventory – as a promising correlate of pain tolerance. Based on sensory estimations of size, the test sorted subjects into 'augmenters' and 'reducers'. Their tolerance to pain – the time difference between their initial perception of painful dolorimetric stimulation and the point at which they asked for it to stop – was then measured. Also measured was the time that subjects were willing to remain in a sensory deprivation tank. By the late 1950s, Petrie and her colleagues announced that 'augmenters' were less tolerant than 'reducers' to painful stimulation and were more tolerant of sensory deprivation.[37] She thus proposed that pain tolerance was linked to perceptual style – a way of experiencing the world.

By entangling pain measurement with personality-defining techniques and methods for measuring the physiological correlates of emotions, psychosomatic researchers differentiated responses to pain along a typology of emotional ways of being. In the 1950s and 1960s, other technologies, some similarly algometric and others based on questionnaires, ethnographic observation and animal conditioning, were deployed to investigate the modulation of pain by personality, upbringing and culture (but, surprisingly, not by gender).[38] Still, the dolorimeter's trajectory through psychosomatic research reveals some of the concrete practices and conditions that produced a personality-dependent (and culturally sensitive) model of pain reactivity. The dolorimeter's very differently shaped trajectory through analgesic research confronted a different kind of 'psychological turn' in pain measurement.

Counting relief: the pursuit of objectivity

The need to evaluate pain-relieving therapies is recognised as a major driver in the history of pain-measuring technologies.[39] This push began with the quest for non-addictive but pain-relieving drugs. In 1929, the Committee on Drug Addiction (CDA) was created under the auspices of the National Research Council. In the early 1930s, molecules produced in the CDA-sponsored chemistry lab at the University of Virginia were screened in animals (using threshold-based methods on algometric principles) at the University of Michigan.[40] By the mid-1930s, two promising drugs were passed on to human subjects to test efficacy but especially addictive liability. Equivocal results led the committee members to reformulate the goals of its quest: for a less addictive rather than non-addictive painkiller. This demanded a finer calculus of analgesic efficacy in humans.

In 1935, the committee members embarked on a purposeful search for objective, quantitative methods of pain evaluation. They consulted with specialists in psychosomatic research (Stanley Cobb, John C. Whitehorn and Walter B. Cannon) about potential physiological indicators of emotional response, but their respondents were sceptical. Threshold-based or algometric methods were also tried both as part of a clinical trial at Pondville Cancer Hospital in Massachusetts and in 'normal subjects' at the Public Health Service's Narcotics Hospital in Lexington, Kentucky. However, results were too variable, suggesting that 'judgemental and interpretive factors were of more significance in producing pain than any particular degree of stimulation'.[41] Meanwhile, the results of clinical trials conducted at Pondville and four other public hospitals in 1937–1939 were considered to have limited value given that they were 'unfortunately' not based 'upon any quantitative measure of analgesic effect'.[42] Lack of clinical authority, coordination and patient volume had made it difficult to enforce protocols controlling conditions of observation (which included the 'blinding' or ignorance of nurses and patients of the drug administered, the use of standardised forms and the pairing of patients with pain of similar origin and intensity).[43]

This is the context in which Wolff approached William Charles White, Chairman of the CDA, at a meeting of the Division of Medical Sciences in 1939. The consistency of the baseline pain threshold produced by the Hardy-Wolff-Goodell dolorimeter, coupled with its sensitivity to drug effects, was exactly what the CDA had been looking for. It was also quick and cheap to use, and applicable to the 'normal' subjects to which the committee had easiest access: the post-addict prisoners of the Lexington narcotics hospital. Arrangements were made for the method to be learned by committee-affiliated researchers and a few trials were run at Lexington.[44] Yet the dolorimeter's enthusiastic reception by the committee was soon interrupted by the termination of Rockefeller funding and the dispersal of its members in wartime research.

The war also brought new German-synthesised analgesic compounds to the USA, namely Demerol (also called Meperidine) in the early 1940s, and Methadone (also called Amidone), reported through government intelligence channels just after the war. Eager to test these new compounds and their derivatives, the pharmaceutical companies Eli Lilly,[45] Hoffman-LaRoche[46] and Whitehall Pharmacal[47] provided grants for specific studies using the dolorimeter, while Smith, Kline & French funded a fellow who worked intensively with other academic researchers on evaluating the dolorimetric method.[48] Their success maintained the popularity of the method throughout the 1940s, thus valorising an experimental pain stripped of its emotional variables.

Yet when the CDA was revived in 1947 as the Committee on Drug Addiction and Narcotics (CDAN), once again on the lookout for a reliable test of pain relief, they rejected the dolorimetric method because of its inapplicability to large numbers of subjects. This criterion was emerging in the 1940s as part of a movement of 'therapeutic reform' to replace expert judgement with impersonal methods and statistical validation of results in judging the value of therapies. The concern was to protect evaluation from bias in general, but particularly the influence of pharmaceutical firms during an explosion in drug development.[49] The choice of Henry K. Beecher's clinical trial method as a promising candidate for committee sponsorship can be linked to these concerns of therapeutic reform. The CDAN's new Chairman, Isaac Starr, Professor of Therapeutic Research at the University of Pennsylvania, formulated the committee's new role as providing 'impartial advice' to government and advising the industry on reliable methods of evaluation in order to avoid the flooding of the market with dangerous drugs.[50] At the time of the CDAN's first meetings in 1947–1948, Beecher was already running a large-scale trial of Methadone and its isomers with Jane Denton at the Massachusetts General Hospital (MGH).[51] The publication of its results in the *Journal of the American Medical Association* was prefaced by an approving statement by the association's Therapeutic Trial Committee (TTC). Active in American therapeutic reform, the TTC had also sent a representative to the CDAN's first meeting in 1947.

If post-war therapeutic reform is now strongly associated with the rise of the Randomised Clinical Trial (RCT), in the late 1940s, clinical trials were not seen as an obvious solution to the problem of accurate and impartial analgesic testing. First of all, trials were expensive and funders had to be convinced they were worthwhile. In 1948, the CDAN invited a group of pharmaceutical firms to pool their resources in its grant programme. At the time, individual firms were funding relatively cheap and rapid analgesic assessments with the dolorimeter or animal tests. Given the prospect of collective funding, representatives admitted that more reliable testing methods were needed and that this would require significant and sustained investment. Their firms provided the entirety of the CDAN's grant budget until 1960.

Second, the method still had to be proven to generate consistent and objective evaluations of analgesia in the absence of individual objective measures of pain. We have seen that the value of previous trials was contested. But Beecher, as he began reporting on his army-funded work, made the remarkable suggestion that the 'safeguards' of clinical evaluation – collecting sufficient volumes of data under controlled conditions (which included standardisation, 'blinding' and the random allocation and comparison of placebos and active treatments) – could in themselves obtain robust measures of relief. In other words, he could state that the raw data of analgesic evaluation represented an irreducibly emotionally saturated and idiosyncratic experience evaluated by real patients' own imprecisely rated pain, and *yet* could be aggregated into a collective and well-controlled indicator of efficacy. Enacting and demonstrating his method's validity, consistency and reproducibility took time, money, coordination and collaboration. Given the durable success of his methodological principles and their role in building consensus not only around the reliability of analgesic evaluation but also the nature of pain as subjective experience, it is worth describing this process in detail.

Counting relief: emotional numbers

The implementation of analgesic clinical trials was facilitated by Beecher's status as Chief of the Anesthesia Service at MGH. His responsibility for post-operative patients gave him ready access to abundant subjects requiring pain relief, as well as the necessary authority over patient care to ensure protocols were followed. The inclusion of placebos, keeping nurses, observers and patients ignorant of when these were administered (a process known as 'blinding'), required the cooperation of physicians and pharmacists, who prepared and labelled the doses with secret codes.

In addition to access and authority, Beecher brought ambition to analgesic evaluation. When discussing his candidacy for sponsorship, members of the CDAN doubted that he would be interested in taking up such a 'dull job' or that the 'touch of genius' required for carrying it through was 'likely be found in the routine operations of a Department of Anesthesia'.[52] Yet Beecher's ambitions for what others did not see as a cutting-edge speciality made him perceive the potential for 'an astonishingly rich harvest' in this admittedly 'painstaking and tedious work'.[53] Upon taking up his post at MGH in 1936, he immediately created a Laboratory of Anesthesia. Along with colleagues, he promoted clinical research as a means of elevating the status of the American anaesthetists (previously subordinated to surgery), and in particular pointed to the professional opportunities offered by analgesic testing and clinical pharmacology of 'subjective responses' more generally.[54] One wonders whether Beecher would have promoted his method so vigorously without the conviction that clinical research could expand his and his protégées' professional opportunities.

If the CDAN tapped into the resources that Beecher brought to analgesic testing, the funding it channelled into his grants allowed him to refine his method. The hiring of full-time observers for collecting pain data may be his most significant methodological innovation. Special observers were more likely than busy clinicians to follow instructions, to accept being 'blinded', to question patients consistently and neutrally, to follow strict schedules and to assure continuity between interviews.[55] Beecher wrote of technicians as a component of his method, listed alongside placebos and randomisation as one of the essential controls in tests of subjective drug effects.[56] He valued the technician for her lack of investment in research or patient care: not knowing, and not caring, about the patient's treatment kept her neutral. Naïve observers with a high turnover rate were the best: Beecher used 'college girls'.[57] Beecher, as well as later grant-holders, spent a large proportion of their CDAN grants on observer salaries.[58] They made the difference between good and bad data: 'the responsibility of testing falls on full-time observers such that objective, quantitative data, rather than clinical impressions, are obtained'.[59]

At a time when the involvement of statisticians in clinical research was still rare, Beecher also spent consultant fees on Frederick Mosteller, a

Harvard colleague, thus helping to launch his career as one of the prominent pioneers of American biostatistics.[60] Statistical expertise was required not only for manipulating results, but also for designing the method so as to produce data that would be optimal for statistical analysis. In the end, Beecher was able to attain a high level of quantitative precision: reporting relief in percentages and plotting dose-effect curves that gave clear distinctions between the efficacy of test and standard analgesics.

If we are to see running successful, consistent clinical trials as teamwork, the most crucial players were probably the patients who served as subjects. Their judgements of pain made up the raw data of analgesic efficacy. The varying ability of subjects to make such judgements worried CDAN researchers. In published reports, Beecher specified only that study subjects should be 'willing, cooperative, undistracted [*sic*]',[61] thus implying that the idiosyncrasies and emotional state of 'ordinary people' would be cancelled out by proper study design. Yet when asked to justify delays, he explained that his studies required 'very careful selection of patients'.[62] Louis Lasagna, who selected non-private preoperative patients for Beecher's studies in the early 1950s, explained that no information about the study was provided (obtaining consent was seen as a source of bias) and that women were excluded because of Beecher's 'stereotype that women, because of the menstrual cycles, have more ups and downs than men do'.[63]

Subjects' performances were also secured by devising appropriate rating scales, adapted to their ability to discriminate between levels of pain intensity or quantities of relief. Beecher's team started with categories of relief: 'none', 'slight', 'moderate' and 'complete'. But these were difficult to turn into meaningful data.[64] Another set of terms – 'one dollar pain', 'seventy-five cent pain' and so on – gave distinctions that were 'not sharp enough'.[65] Finally, they settled on two categories: relief and no relief, the first being defined as 'the disappearance of "most" or "more than half" of the pain'. Anything less was counted as 'no relief'.[66]

In the early 1950s, Beecher began to turn away from the immediate practical details of how to run analgesic clinical trials. Using CDAN grants, he investigated phenomena such as conditioning, suggestion and placebo effects. These studies initially aimed to identify sources of error in analgesic testing, but broadened to questions of how pain and relief were experienced. These studies, and the more minor methodological tinkering Beecher engaged in, were made possible by the CDAN's mediation of industry funds. Starr insisted that the research programme should not 'degenerate into a simple matter of clinical testing' (serving industry) and should support 'fundamental research' bearing on methodological issues and a broader understanding of analgesia.[67] Beecher and his colleagues hoped tests of subjects' personalities, susceptibility to suggestion and conditioning, ability to manipulate different types of measurement instruments and reaction to being asked for consent would help tighten their control over experimental conditions. If most of these studies had no concrete application, they treated and thus helped understand pain as the experience of thinking and feeling individuals.

To sum up, the successful implementation of the analgesic clinical trial, which provided evidence for the measurability of analgesia using subjective reports of pain experienced in its 'natural' and thus emotionally saturated habitat, was made possible by the following factors: Beecher's access, authority and ambitions as a clinical chief of service in a low but rising prestige specialty; the CDAN's ability to attract, combine and maintain 'scientific autonomy' in channelling funds from the pharmaceutical industry; the hiring of full-time observers and of a statistician's services; the time and freedom to engage in methodological tinkering and investigations; and finally the cooperation of patients whose subjectivity was monitored and investigated. In addition, the diffusion of this method to multiple sites was fairly slow and tightly controlled by CDAN sponsorship and personal connections among researchers, thus improving the chances of reproducing its results.

Without the data produced by and for the application of Beecher's trial methods, would he have been able to state with such confidence that pain was indivisible at the level of experience? He chided previous pain measurers for attempting to split off the sensation of pain from the reaction to it, arguing that reaction preceded conscious awareness of pain. As bundled sensation and reaction, pain was 'never alike for any two individuals and, indeed, with the passing of time and accumulation of life experience, is never exactly the same for the same individual from one time to another'.[68] The subjectivity of pain was not new, but the suggestion that it could be reliably measured without being reduced to something less personal and more predictable was new indeed.

Pain as an indivisible experience not only justified the analgesic clinical trial but was also justified by it. The regularities of collective pain relief in the 'bird's eye' view protected individual pain from interference. This variability at the individual level, its openness to emotion, interpretation, memory, attention and so on was declared beyond control, constitutive of the experience of pain itself. Thus, we can see Beecher's work in the late 1940s and 1950s as an important episode in the historical process of authorising emotion as a component of anything worth calling 'pain'.

Notes

1. B. Noble, D. Clark, M. Meldrum, H. Ten Have, J. Seymour, M. Winslow and S. Paz, 'The Measurement of Pain, 1945–2000', *Journal of Pain and Symptom Management*, 29(1) (2005): 14–21.
2. Lorraine J. Daston and Peter L. Galison, *Objectivity* (New York: Zone Books, 2007).
3. Gail A. Hornstein, 'Quantifying Psychological Phenomena: Debates, Dilemmas, and Implications', in J.G. Morawski (ed.), *The Rise of Experimentation in American Psychology* (New Haven: Yale University Press, 1988), 1–34.
4. Edwin G. Boring, *Sensation and Perception in the History of Psychology* (New York: Appleton-Century Company, 1942), 467.
5. Roselyne Rey, *The History of Pain* (Cambridge, MA: Harvard University Press, 1998), 215–16.
6. For example, B. Berthold Wolff, 'Laboratory Methods of Pain Measurement', in Ronald Melzack (ed.), *Pain Measurement and Assessment* (New York: Raven Press, 1983), 7.
7. Karl M. Dallenbach, 'Pain: History and Present Status', *American Journal of Psychology*, 52 (1939): 331–47.
8. Karl Danziger, *Constructing the Subject* (New York: Cambridge University Press, 1990), 5.
9. Deborah J. Coon, 'Standardizing the Subject: Experimental Psychologists, Introspection, and the Quest for a Technoscientific Ideal', *Technology and Culture*, 34 (1993): 757–83; Ruth Benschop and Douwe Draaisma, 'In Pursuit of Precision: The Calibration of Minds and Machines in Late Nineteenth Century Psychology', *Annals of Science*, 57 (2000): 1–25.
10. For example, Graham Richards, *'Race', Racism and Psychology: Towards a Reflexive History* (London: Routledge, 1997).
11. *Ibid.*, 44.
12. A.C. Haddon (ed.), *Reports of the Cambridge Anthropological Expedition to Torres Straits*, vol. 2 (Cambridge University Press, 1901–35), 195.
13. G. Lombroso-Ferrero (ed.), *Criminal Man According to the Classification of Cesare Lombroso* (Montclair, NJ: Patterson Smith, 1972), 246–9; C. Lombroso, *Nouvelles Recherches de Psychiatrie et d'Anthropologie Criminelle* (Paris: F. Alcan, 1892), 88.
14. Mary Gibson, 'On the Insensitivity of Women: Science and the Woman Question in Liberal Italy, 1890–1910', *Journal of Women's History*, 2 (1990): 15, 35n10.
15. Cited in *ibid.*, 17–18.
16. Arthur MacDonald, *Hearing on the Bill (H. R. 14798) to Establish a Laboratory for the Study of the Criminal, Pauper, and Defective Classes, with a Bibliography* (Washington DC: Government Printing Office, 1902); Arthur MacDonald, *A Plan for the Study of Man with Reference to Bills to Establish a Laboratory for the Study of the Criminal, Pauper and Defective Classes* (Washington DC: Government Printing Office, 1902).
17. MacDonald, *Hearing on the Bill*, 41.
18. James B. Gilbert, 'Anthropometrics in the U.S. Bureau of Education: The Case of Arthur MacDonald's "Laboratory"', *History of Education Quarterly*, 17 (1977): 169–95.

19. Rob Boddice, 'The Manly Mind? Revisiting the Victorian "Sex in Brain" Debate', *Gender & History*, 23 (2011): 321–40; Lucy Bending, *The Representation of Bodily Pain in Late Nineteenth-Century English Culture* (Oxford: Clarendon Press, 2000).
20. James D. Hardy, Harold. G. Wolff and Helen Goodell, 'Studies on Pain. A New Method for Measuring Pain Threshold: Observations on Spatial Summation of Pain', *Journal of Clinical Investigation*, 19 (1940): 649–57.
21. 'Brothers under the Skin: We All Feel Pain in Same Way, National Academy is Told', *Newsweek*, 6 May 1940.
22. Anonymous, 'Obituary', Wolff Papers, Box 1, folder 1, Medical Center Archives of New York-Presbyterian/Weill Cornell, New York.
23. William P. Chapman, 'Measurements of Pain Sensitivity in Normal Control Subjects and in Psychoneurotic Patients', *Psychosomatic Medicine*, 6 (1944): 252–7.
24. Richard B. Malmo, Charles Shagass and J.F. Davis, 'Pain as a Standardized Stimulus for Eliciting Differential Physiological Responses in Anxiety', *American Psychologist*, 2 (1947): 344; Richard B. Malmo, Charles Shagass, J.F. Davis, R.A. Cleghorn, B.F. Graham and A. Joan Goodman, 'Standardized Pain Stimulation as Controlled Stress in Physiological Studies of Psychoneurosis', *Science*, 108 (1948): 509–11. This institute was located in Montreal, Canada, but the researchers participated in American networks of psychosomatic research.
25. Asenath Petrie, W. Collins and P. Solomon, 'Pain Sensitivity, Sensory Deprivation, and Susceptibility to Satiation', *Science*, 128 (1958): 1431–3.
26. Robert Powell, 'Healing and Wholeness: Helen Flanders Dunbar (1902–59) and the Extra-Medical Origins of the American Psychosomatic Movement, 1906–36' (PhD thesis, Department of History, Duke University, 1974); Otniel E. Dror, 'The Affect of Experiment: The Turn to Emotions in Anglo-American Physiology, 1900–1940', *Isis*, 90 (1999): 205–37; Jack D. Pressman, 'Human Understanding: Psychosomatic Medicine and the Mission of the Rockefeller Foundation', in Christopher Lawrence and George Weisz (eds), *Greater Than the Parts: Holism in Biomedicine 1920–1950* (Oxford University Press, 1998), 189–208.
27. Franz Alexander, *Psychosomatic Medicine: Its Principles and Applications* (New York: Norton, 1950); E. Weiss and O. English, *Psychosomatic Medicine: The Clinical Application of Psychopathology to General Medical Problems* (Philadelphia and London: W.B. Saunders, 1943).
28. Asenath Petrie, *Personality and the Frontal Lobes: An Investigation of the Psychological Effects of Different Types of Leucotomy* (Philadelphia: Blakiston, 1952); M. Vidor, 'Personality Changes Following Prefrontal Leucotomy as Reflected by the MMPI and the Results of Psychometric Testing', *Journal of Mental Science*, 97 (1951): 159–73; J. Charney McKinley and Starke Hathaway, 'The Identification and Measurement of the Psychoneuroses in Medical Practice: Minnesota Multiphasic Personality Inventory', *JAMA*, 122 (1943): 161–7; Petrie, *Personality*, 6.
29. James W. Watts and Walter Freeman, 'Frontal Lobotomy in the Treatment of Unbearable Pain', *Proceedings of the Association for Research in Nervous and Mental Diseases*, 27 (1948): 715–22.
30. Watts and Freeman, 'Frontal Lobotomy'; James D. Hardy, Harold G. Wolff and Helen Goodell, *Pain Sensations and Reactions* (New York: Hafner, 1952).

31. William P. Chapman and Chester M. Jones, 'Variations in Cutaneous and Visceral Pain Sensitivity in Normal Subjects', *Journal of Clinical Investigation*, 23 (1944): 81–91; William P. Chapman, J.E. Finesinger, Chester M. Jones and Stanley Cobb, 'Measurements of Pain Sensitivity in Patients with Psychoneurosis', *Archives of Neurology and Psychiatry*, 57 (1947): 321–31; William P. Chapman, A.S. Rose and H.C. Solomon, 'Measurements of Heat Stimulus Producing Motor Withdrawal Reaction in Patients Following Frontal Lobotomy', *Proceedings of the Association for Research on Nervous and Mental Disorders*, 27 (1948): 754–68.

32. William P. Chapman, M.E. Cohen and Stanley Cobb, 'Measurements Related to Pain in Neurocirculatory Asthenia, Anxiety Neurosis, or Effort Syndrome: Levels of Heat Stimulus Perceived as Painful and Producing Wince and Withdrawal Reactions', *Journal of Clinical Investigation*, 25 (1946): 890–6. Cobb's psychiatric service was among those supported by funds from the Rockefeller Foundation for psychosomatic research (Pressman, 'Human Understanding').

33. Malmo, Shagass and Davis, 'Pain as a Standardized Stimulus', 344. Malmo *et al.*, 'Standardized Pain Stimulation', 509–11.

34. Dror, 'The Affect of Experiment'.

35. Petrie, *Personality*, 16–21.

36. Petrie, *Personality*.

37. Petrie, Collins and Solomon, 'Pain Sensitivity'; Asenath Petrie, *Individuality in Pain and Suffering* (University of Chicago Press, 1967).

38. Ronald Melzack, 'The Effects of Early Experience on the Emotional Responses to Pain' (PhD thesis, McGill University, 1954); Donald V. Petrovich, 'The Pain Apperception Test: Psychological Correlates of Pain Perception', *Journal of Clinical Psychology*, 14 (1958): 367–374; Mark Zborowski, *People in Pain* (San Francisco: Jossey-Bass, Inc. 1969).

39. Noble *et al.*, 'The Measurement of Pain'.

40. Caroline J. Acker, *Creating the American Junkie: Addiction Research in the Classic Era of Narcotic Control* (Baltimore: Johns Hopkins University Press, 2002), 62–97.

41. E. Williams, 'A Quantitative Measure of Analgesia (Summary of Work Done at Lexington in Past Year, November 12, 1938' in L. Kolb to White, 12 November 1938, Projects: Development of Nonaddictive Analgesics: Clinical studies: PHS Hospital: Lexington General, Committee on Drug Addiction (CDA), National Academies of Science Archives (NASA).

42. 'Report on Work on Analgesia at the University of Michigan, October 6, 1937', Projects: Development of Nonaddictive Analgesics: Clinical Studies: University of Michigan Hospital, CDA, NASA.

43. 'Memorandum Number 2, June 18, 1936', Projects: Development of Nonaddictive Analgesics: Clinical Studies: Pondville Hospital, CDA, NASA; Eddy to Metcalfe, November 18, 1938, Projects: Development of Nonaddictive Analgesics: Clinical Studies: Walter Reed General Hospital, CDA, NASA; Lyndon Lee to Eddy, 21 November 1938, Projects: Development of Nonaddictive Analgesics: Clinical Studies: Pondville Hospital, CDA, NASA; Lee to White, 8 December 1938, Projects: Development of Nonaddictive Analgesics: Clinical Studies: Pondville Hospital, CDA, NASA; Denney to Eddy, 13 November 1939, Projects: Development of Nonaddictive Analgesics: Clinical Studies: Marine Hospital, 1938–1939, CDA, NASA.

44. 'Report to the Surgeon General on His Visit to Wolff's Laboratory by H.L. Andrews, July 28, 1939', Projects: Development of Nonaddictive Analgesics: Clinical Studies: PHS Hospital: Lexington General, CDA, NASA.

45. C.C. Scott, E.B. Robbins, and K.K. Chen, 'Pharmacologic Comparison of the Optical Isomers of Methadon', *Journal of Pharmacology and Experimental Therapeutics*, 93 (1948): 147–56.

46. D. Slaughter and F.T. Wright, 'A Modification of the Hardy-Wolff-Goodell Pain-Threshold Apparatus', *Current Research in Anesthesia and Analgesia*, 23 (1944): 115–19; E.G. Gross, H.L. Holland and F.W. Schueler, 'Human Studies on Analgesic Piperidine Derivatives', *Journal of Applied Physiology*, 1 (1948): 298–303.

47. F.B. Flinn and A.S. Chaikelis, 'An Improved Instrument for the Determination of Changes in the Pain Threshold Caused by Drugs', *American Journal of Psychiatry*, 103 (1946–7): 349–50.

48. Robert. R. Sonnenschein and A.C. Ivy, 'Failure of Oral Antipyretic Drugs to Alter Normal Human Pain Thresholds', *Journal of Pharmacology*, 97 (1949): 308–13.

49. Harry M. Marks, *The Progress of Experiment: Science and Therapeutic Reform in the United States, 1900–1990* (Cambridge University Press, 1997).

50. 'Minutes of the 1st Meeting', *Bulletin of the Committee on Drug Addiction and Narcotics* (1947): 6.

51. It was funded by the Medical Research and Development Board of the US Army.

52. Starr to Eddy, 13 July 1949, Box 2: General, 1947–June 1959, CDAN, NASA; Philip Owen (NRC) to Starr, 20 July 1949, Box 2: General, 1947–June 1959, CDAN, NASA.

53. Henry K. Beecher, 'Experimental Pharmacology and Measurement of the Subjective Response', *Science*, 116 (1952): 162.

54. R. Charles Adams, 'Clinical Research in Anesthesiology', *Anesthesiology*, 11 (1950): 178–84; 'Editorial: Clinical Investigation', *Anesthesiology*, 12 (1951): 114–18; N.M. Greene, *Anesthesiology and the University* (Philadelphia: J.B. Lippincott Company, 1975), 45–8; Henry K. Beecher, 'Pain and Some of the Factors that Modify it', *Anesthesiology*, 12 (1951): 633–41.

55. H.M. Marks, 'Trust and Mistrust in the Marketplace: Statistics and Clinical Research, 1945–1960', *History of Science*, 38 (2000): 343–55.

56. Beecher, 'Experimental Pharmacology', 160.

57. Henry K. Beecher, 'Studies on Narcotics: Annual Report', *Bulletin of the Committee on Drug Addiction and Narcotics* (1955): 1073.

58. We see this in applications for support to CDAN. For example, in 1952, Beecher asked for $5,980 to pay technicians out of a total $16,679 budget (part of which went to a separate study not employing observers, so the salary proportion in the cost of the clinical trial was higher). 'Annual Report and Application for Renewal of Support to the Committee on Narcotics and Drug Addiction of the National Research Council, 21 January 1952', *Bulletin of the Committee on Drug Addiction and Narcotics* (1952): 225.

59. 'Minutes of the 4th Meeting', *Bulletin of the Committee on Drug Addiction and Narcotics* (1948): 61. Beecher presented preliminary results of his army-funded trials at this meeting.

60. Marks, *The Progress of Experiment*, 129–63.

61. Beecher, 'Experimental Pharmacology', 160.
62. 'Minutes of the 8th Meeting', *Bulletin of the Committee on Drug Addiction and Narcotics* (1951): 192–3.
63. Marcia Meldrum, 'Oral History Interview with Louis Lasagna', 8 September 1995, MS C 127.19, John C. Liebeskind History of Pain Collection, HSCD, Darling Library, UCLA, 4–5, 35; Beecher, 'Experimental Pharmacology', 160.
64. Henry K. Beecher, 'A Method for Measuring Pain in Man', *Bulletin of the Committee on Drug Addiction and Narcotics* (1953): 649.
65. *Ibid.*, 658.
66. *Ibid.*, 649.
67. 'Minutes of the 11th Meeting', *Bulletin of the Committee on Drug Addiction and Narcotics* (1953): 388.
68. Henry K. Beecher, 'Limiting Factors in Experimental Pain', *Journal of Chronic Diseases*, 4 (1956): 11–21.

8
Killing Pain? Aspirin, Emotion and Subjectivity

Sheena Culley

In 1998, Damien Hirst opened *Pharmacy*, a trendy concept restaurant in collaboration with Matthew Freud, Liam Carson and Jonathan Kennedy, respected figures in the food industry. The interior was designed by Hirst, coinciding with his installation of the same name at Tate Britain. Both works featured contemporary medicines and their packaging as their key materials. *Pharmacy*'s restaurant interior included aspirin bar stools and an aspirin print bar illuminated by a light box. The restaurant, like many of Hirst's projects, enjoyed much publicity. The Royal Pharmaceutical Society accused *Pharmacy* of being misleading to the public, causing its creators to change the name to *Pharmacy Restaurant & Bar*. The turning point, according to Hirst, was a customer asking for an aspirin.[1] Hirst's installations, if only for a short time, propelled the mundane aspirin into the public arena, alongside other quotidian pills and potions.

Despite the tendency for us to overlook aspirin in developed, industrialised countries, the history of this chalky, circular, white pill has received some interest in recent years. Aspirin belongs to a group of medicines known as non-steroid anti-inflammatory drugs (NSAIDs), of which aspirin was the first to be commercially successful. Literature on the history of aspirin tends to focus on two main qualities of the drug: first, it is an antipyretic, with the ability to reduce the body temperature of a febrile person; and, second, it is an analgesic, a substance with the ability to alleviate pain. The type of pain we associate with aspirin and other over-the-counter painkillers such as ibuprofen and paracetamol could be classed as 'physical' pain. Take, for example, a box of Nurofen (a brand of ibuprofen, a more recent NSAID). The manufacturers claim that it offers 'rapid relief from: headache, migraine, backache, period pain, dental pain, neuralgia, rheumatic pain, muscular pain, feverishness, and cold and flu symptoms'.[2]

Two recent historical studies of aspirin, Diarmuid Jeffreys' *Aspirin: The Remarkable Story of a Wonder Drug* and Jan McTavish's *Pain and Profits: The History of the Headache and its Remedies in America*, concentrate on the development of aspirin as an antipyretic and analgesic for physical pain.[3] Both are important texts, bringing to life the history of this banal medicine. However, to write the history of an object from the standpoint of its current or even intended use potentially conceals other ways in which it was consumed. Whereas Jeffreys constructs a detailed history of the characters involved in the commercial development of the drug, McTavish focuses more closely on the cultural connotations of the headache in North American culture, demonstrating that people's understanding of such a phenomenon has changed over time. Today, a headache is used to describe a trouble or annoyance, akin to the saying 'a pain in the neck', which marks a trivialisation of a sort of pain that can be anaesthetised or 'killed' by the consumption of over-the-counter medication.[4] McTavish's work also asks us to challenge the conception of a headache as 'physical' pain. Headaches, she states, have been seen since the time of humoral medicine as symptoms rather than diseases themselves, understood as a means to communicate other underlying problems. Disease was not viewed as an invader of the body, as we often see it today, but as an imbalance of the humours.[5] As a result, the headache of the nineteenth century or its underlying cause could be viewed as a 'reaction to diet, the weather, emotions, bad habits, new activities and so forth'.[6] Imbalance of the humours encompasses the 'physical' and 'emotional'. The fact that this view of the headache did not disappear with the advent of aspirin invites the questioning of the relationship between over-the-counter analgesics and a pain that was not entirely corporeal.

Although it would be an oversimplification to state that the headache, in the West today, is completely devoid of emotional connotations (after all, we refer to 'stress headaches' and 'tension headaches'), there has no doubt been a shift that has resulted in a distinction between 'physical' and 'emotional' pain that can be interpreted as more than a Cartesian split. This is the key problem for David B. Morris in his seminal work *The Culture of Pain*. Morris states that 'our culture – the modern, Western, industrial, technocratic world – has succeeded in persuading us that pain is simply a medical problem'.[7] The medicalisation of pain was of course inseparable from developments in anaesthesia. Surgical anaesthesia was discovered through the recreational use of nitrous oxides in 1846 and aspirin first synthesised from salicylic acid in 1899. Morris goes as far as to say that 1899 signified the 'advent of modernism', marking a threshold between the 'preanaesthetic' and 'anaesthetic' modern world, a transition that altered human experience.[8] He also states:

> Probably no other drug – not even such modern favourites as Valium or cocaine – has established itself so firmly in our culture as aspirin. Yet aspirin is far more than our most common over-the-counter analgesic. It is an emblem of our immense faith in chemical assaults on pain.[9]

Significantly, Morris notes that the invention of modern anaesthesia and our ability to 'kill' pain did not lead to its death: 'The pills in a sense just make things worse.'[10] It is thus important to ask how aspirin and its successors have shaped 'cultures of pain'. This can add to our understanding of pain in ways that medical knowledge cannot, and we can glean fragments of an understanding of a certain modern pain by analysing the way aspirin was historically consumed and sold to the public.[11] In order to understand modern pain and its relationship to subjectivity, I draw upon Foucault's *The History of Sexuality* to interrogate the culture of over-the-counter analgesics. In Volume 1 of *The History of Sexuality*, Foucault, by disproving the 'repressive hypothesis' of sex, shows us how the perceived absence of sex illuminates its importance in culture and the role it plays in shaping subjectivity.[12] Victorian sexuality was not 'prohibited or barred or masked'.[13] Instead, Foucault argues, sexuality entered a 'discursive explosion', which saw a diffusion of sexuality into public health and hygiene, population control and normalising definitions. These processes, all contributing to a 'science of sexuality', defined sexuality for Foucault as a matter of power relations, helping to define the historical subject:

> Sexuality must not be thought of as a kind of natural given which power tries to hold in check, or as an obscure domain which knowledge tries gradually to uncover. It is the name that can be given to a historical construct.[14]

The History of Sexuality is a useful foundation for this study for two reasons. First, the perceived 'disappearance' of pain in our anaesthetic age, as with the perceived repression of sexuality in Foucault's text, provides an insight as to how a particular type of modern pain might be constructed by its perceived absence. Second, drawing on examples of advertisements for aspirin and other over-the-counter painkillers, we see evidence of some of the power relations that Foucault uncovers about female sexuality, in particular the hysterisation of the female body and the role of medicine in producing discourses of sex through 'nervous disorders'.[15] The image of the neurotic, hysterical woman in advertisements for aspirin and related products from 1910 to 1959 from the USA, Britain and Australia is important in defining a certain type of modern, female pain that is both 'emotional' and 'physical', and thus a certain type of modern female subjectivity. Without advocating an abstinence from painkillers, the political implications of our 'immense faith' in over-the-counter analgesics need to be addressed.

The origins of aspirin

Although the development of aspirin as a commercial product is bound up with modernity, its active chemical, salicylic acid, has been in existence for thousands of years. This substance is naturally occurring in the bark of specific trees, including the willow. It was said to be recommended as a painkiller during childbirth and as a fever reducer by Hippocrates, and in AD 30 by the Roman physician Celcus to reduce swelling. It was also known to be used by the Ancient Egyptians.[16] In 1753, Reverend Edward Stone of Chipping Norton made an important discovery about the connection between willow bark and the alleviation of fever, although it was dismissed by the medical establishment. Despite these discoveries, salicylic acid's potential was under-recognised in medicine until the nineteenth century.

Synthetic drugs for the relief of pain were discovered by accident through their use as antipyretics. Salicylic acid was first synthesised by Hermann Kolbe, Professor of Chemistry at Leipzig University, in 1874. He created salicylic acid from phenol (the drug naturally occurs in willow trees such as Salix alba).[17] It lowered temperatures and reduced swelling and inflammation, but had terrible side-effects on the stomach in its salt form, sodium salicylate.[18] Thus, by the 1880s, it had lost its popularity. In 1882, Otto Fischer synthesised a drug that he called kairin, which was manufactured by the German firm Hoechst. This foreshadowed the important synthetic antipyretics that were used before the invention of aspirin – antipyrine, antifebrin and phenacetin – which were all invented between 1884 and 1887, with the intention to alleviate fever.[19]

McTavish states that it was the patients themselves who were the first to discover the analgesic effects of these early antipyretics, who in turn informed their doctors. John Blake White, an American doctor, was documented to have observed such effects on his patients, stating that that 15 grains of antipyrine given in a single dose 'promptly relieves the symptom of headache whenever present, whether resulting from disordered digestion, disturbances of the menstrual function, loss of sleep, undue mental effort, or even that associated with the dreaded uraemia'.[20] However, these drugs were not commonly prescribed for the complaints described above. Headaches continued to be seen as the symptom of an imbalance, adhering to the view of humoral medicine. Thus, recommendations for a headache as late as 1916, 17 years after the invention of aspirin, included 'a visit to the seashore or mountains, or massage, restriction of sexual activity, or a long stay at a sanatorium'.[21]

It was finally in 1889 when aspirin was produced by three young scientists, Felix Hoffman, Heinrich Dreser and Arthur Eichengrün, working at *Farbenfabriken vormals Friedrich Bayer* (the manufacturing company formally known and hereafter be referred to as Bayer), which started life as a dye-manufacturer in Elberfeld, north-west Germany. The drug was initially marketed as a fever reducer and anti-inflammatory which 'showed some promise as an analgesic'.[22] After initial success, Bayer realised that it could exploit sales if it patented the drug. It was unsuccessful in Britain, as the patent was for the manufacturing of the chemical, acetylsalicylic acid, but secured success in the USA and hence began an aggressive marketing campaign, not to the general public but to the medical trade.[23] Bayer pushed the brand name 'aspirin', chosen for its simplicity so that doctors would prescribe it over the generic chemical substance. Aspirin became hugely profitable for Bayer and enabled its development from a middle-sized company to one of Germany's largest.[24] Aspirin then started to be marketed to the public, appearing in newspaper advertisements in publications such as the *New York Times* from July 1916 and the *Manchester Guardian* from 1910.[25] The period mainly spanning the two World Wars is referred to as 'the aspirin age', and it is during this time that aspirin became globally popular.[26] Its value was greatly appreciated by many during the Great Influenza Pandemic of 1918–1919. Aspirin was always affordable, although its mark-up made it hugely profitable. Due to its inexpensive and abundant availability, it was used by people from all backgrounds. Following the First World War, Bayer lost its US assets and the aspirin trademark, seeing a further democratisation of the drug. This freedom led many companies into increased competition, and new ranges of analgesic products came onto the market. These included Dispirin, a soluble aspirin, a British product introduced in 1948, and Panadol, a brand of paracetamol, introduced in 1956, also in Britain.[27]

Advertising a cure-all

An early advertisement for Genaspirin (a brand of aspirin) in the *Manchester Guardian* in 1919 contains the following copy:

Sleeping badly? – then get a free sample of Genaspirin. Swallow two of these harmless little tablets – disintegrated in water – before 'turning in'. They quieten the excited brain – calm the throbbing nerves – and so predispose you to healthy natural sleep.[28]

The 'soothing, sedative effect' that this product promises is surprising to us today, as we do not think of aspirin as a sedative. However, the use of this aspirin product can be compared to the opiates that preceded it. Opium was used not only as an analgesic, but also for a variety of ailments, for example, as a sedative for children, medication for insomnia and in 'tranquillisers for the insane'. It was also the leading drug for nervous exhaustion, one of the terms relating to neurasthenia: a modern, all-encompassing disease of the nerves.[29] It was therefore not unusual to expect a painkiller to have emotional and physical uses in the late nineteenth and early twentieth centuries.

It is perhaps even more intriguing to contrast this example with an excerpt from George Orwell's *The Road to Wigan Pier* (1937). Orwell comments on the consumption of aspirin when observing the eating habits of the working classes: 'a cup of tea or even an aspirin is much better as a temporary stimulant than a crust of brown bread'.[30] Not only did aspirin have uses other than killing corporeal pain, but the same medicine was used both as a sedative and as a stimulant. However, the changing and contradictory uses of substances – from food to tea, coffee, alcohol and medicines – was not uncommon during the period of modernity. We could compare the consumption of aspirin to that of coffee. By the seventeenth century, it was hailed as a 'panacea', able to fortify the liver and gall bladder, to cure colic, to soothe an upset stomach, and paradoxically both to whet and decrease the appetite, and to keep one awake as well as to induce sleep.[31] Aspirin was similarly viewed as a panacea. Seemingly contradictory uses highlight the need to understand aspirin in terms of its social and cultural significance rather than by its strictly physiological effects, 'proven' by the medical profession. As Deborah Lupton states, the physical effects of substances are difficult to separate from the 'cultural expectation' that accompanies them.[32] What we believe a substance to do is crucial to our experience of it. Moreover, it would seem that aspirin was intentionally sold as a cure-all. The initial advertisements for Bayer aspirin to physicians pitched the drug as a remedy for 'baby colic, colds, influenza, joint pain and other ailments – even as a general pick-me-up', much like Coca-Cola.[33] What was the role of advertising in selling these expectations to the aspirin-consuming public, with particular focus on the female consumer?

Aspirin, hysteria and nervous disorders

As well as the low price of aspirin, advertising played a central role in the rise of the drug. It must be emphasised that aspirin was always a commercial product, invented by researchers at a commercial firm. Advertising was not new at this time in the case of self-prescribed medication. Nostrums (also known as patent medicines or proprietaries), popular remedies in the nineteenth century, relied heavily on newspaper advertising. Nostrums originated in Britain and the potions often owed their ingredients to herbal medical practices. One could visit a range of outlets, including 'postmasters, goldsmiths, grocers, hairdressers, tailors, painters, booksellers, cork cutters … and physicians' to obtain such remedies.[34] Unlike medicines obtained from a pharmacy at the time, nostrums were pre-packaged in bottle sizes appropriate for domestic consumption. The manufacturer did not have to state the ingredients of the preparations, meaning that grand claims could be invented around these mysterious medicines.

The popularity of nostrums began to rise in the USA at a time where doctors were hard to come by and feared (there was no professional medical body until 1849). One notable nostrum propelled women's pain into the medicine market: Lydia E. Pinkham's Vegetable Compound, which was advertised as offering 'the surest remedy for all the painful ills and disorders suffered by women everywhere'.[35] Pinkham began making the remedy at home and first sold it as a commercial product in 1873, and she is reported to have become one of America's first women millionaires.[36] Jeffreys is sceptical of the product, stating that its popularity 'owed much to the fact that its most active ingredient was alcohol'.[37] Whether or not this nostrum functioned as an analgesic is only part of the concern, as what is overlooked, as stated by Susan Strasser, is that it 'gave opportunities for women to relate to a commercial character they might trust'.[38] Strasser is referring to the personal appeal of the bottle, which featured Lydia Pinkham's portrait, and the 'agony aunt' service she provided to her customers. The packaging and advertising were central to the success of the product, giving women hope of pain relief, if not relief itself, from a variety of complaints, including pain during menstruation. Thus, before the introduction of aspirin, there was both demand for pain relief for women and the expectation, or at least the hope, that it could be sought in the form of an over-the-counter remedy.

In the history of aspirin advertising we see a common theme of the female target customer. It is not the benefit to women that is the focus here, but rather how a social construction of the feminine effected the language of advertising and presented a gendered understanding of pain. Two conditions, in their nineteenth-century forms, helped to create a specific female pain that facilitated targeting by manufacturers of aspirin and other painkillers: nervous disorders and hysteria.

The broad category of nervous pain and illness encapsulated the dominant discourse of emotional suffering in the Western world of the eighteenth and nineteenth centuries. Nervous conditions can be seen as a precursor to depression, although the term has a complex history, and it is important to note that at this time there was no pathology of depression and therefore no such medicine as an antidepressant, at least as we know it today. Nervous conditions were initially confined to the wealthy, and it was luxuries of the day such as tea, coffee, spicy foods and general over-indulgence and excess that were said to have upset the delicate disposition of the upper classes. In George Cheyne's *The English Malady* (1733), nervous illnesses included lowness of spirits, hysteria and hypochondria, which were diseases of the aristocracy.[39] Although the diagnosis of nervous conditions was initially confined to the wealthier section of society, they were democratised along with the luxury that industrialisation brought. During the fast-paced speed of industrialisation, the disease of neurasthenia was documented by George Beard in his book *American Nervousness, its Causes and Consequences* (1881), where he attributed the new malady to modern civilisation. Neurasthenia was a disease with over 70 mental and physical symptoms, and nervous disorders created 'an overly inscribable body, one too easily written upon by the stimulus of its day to day experience'.[40] 'Shocks' from a variety of modern technologies played a part in this process, from printing presses to railways. Even those in sedentary occupations such as bank clerks and salespeople were vulnerable to such a disease in the nineteenth century.[41]

Although both men and women could suffer from the distinctly modern disease of neurasthenia, there has been discussion surrounding the gendering of the condition. For example, due to the diagnosis resulting from overwork, it could be seen as a male condition, but due to beliefs about the inherent weakness of the female nervous system, the argument swings the other way.[42] Elaine Showalter reads neurasthenia as a nineteenth-century manifestation of hysteria. Hysteria is derived from the Greek 'hystera', the word for uterus. The Ancient Greeks believed that the uterus migrated around the body, causing myriad symptoms. According to Showalter, hysteria can be considered to be 'mimetic' – that is, it 'culturally mimics expressions of distress'.[43] To accept this argument is not to reduce neurasthenia to hysteria, but to show that there existed an hysterical undertone to women's nervous disorders at the time. In addition, Foucault refers to the 'hysterisation of women's bodies', part of the nineteenth-century administration of sexuality, which had a central role to play in defining female sexuality in terms of nervousness.[44] Two famous doctors helped to define the nervous hysteria of the late nineteenth and early twentieth centuries: Jean-Martin Charcot and Sigmund Freud. Charcot (1825–1893) believed that hysteria was a neurological condition brought about by heredi-tary features of the nervous system. His patients included men and women at the Salpêtrière (there were 90 male hysterics in total, but, it is thought, ten times this number of women), and as Mark Micale observes, his work intentionally challenged hysteria as an exclusively female malady. Although these male hysterics were not effeminate but 'vigorous' artisans, curiously Charcot attributed hysteria to ovarian sen-sitivity. In men he would attempt to find parallel 'hysterogenic' bodily regions that he claimed to have discovered in the female hysteric.[45] Freud's development of Charcot's studies did nothing to relieve hysteria of its mysterious nature. Whereas Charcot had focused on seizures as the primary corporeal symptom of hysteria, Freud argued that hysteria was more subtle and characterised by 'everyday symptoms of *petit hysté-rie*: coughs, limps, headaches, loss of voice', and also insomnia, anxiety and other physical pains.[46] His patients, such as Anna O. and Dora, were typically female, and their hysteria brought about by trauma. The importance of sexual difference facilitated the classification of vari-ous pains under the general category of hysteria, separating male and female pain.[47] As David B. Morris states, due to the history of hysteria and its gendered dimension, the cultural context of female pain remains 'elusive and harder to see'.[48]

A gendered version of nervous pain is exemplified in advertisements for Coca-Cola, which started life as a 'nerve tonic' in 1886, three years prior to the synthesis of aspirin. Coca-Cola's founder, John Pemberton, was fascinated by coca leaves and 'Vin Mariani', a concoction of wine and coca leaf, which become popular in the USA in the 1880s. The original Coca-Cola contained wine, cocaine and caffeine, and was essentially a nostrum like the others of its time. The original advertisement promised that Coca-Cola was a 'cure for all nervous affections – Sick Head-Ache, Neuralgia, Hysteria, melancholy, etc'. However, from an advertisement from 1905, we see that these conditions for men might have been brought about by their profession – the terms 'business', 'professional', 'students', 'wheelmen' and 'athletes' are used. The male consumer of Coca-Cola was seeking to 'relieve mental and physical exhaustion'; the cause of his nervous affections was clear. However, the same Coca-Cola, a favourite drink of ladies, was to be consumed by women because they were 'thirsty, weary and despondent'.[49] Despondency, suggestive of despair and hopelessness, set women's nervousness apart from men's. Rather than a symptom of overwork, as it was for men, a nervous disposition had no specific cause. Even Lydia Pinkham's Vegetable Compound was not free from targeting its specifically female remedy to the hysterical woman. The headline to one advertisement explicitly evoked 'these hysterical women', with the copy continuing 'tired all the time ... overwrought ... nerves strung to breaking point. Constant headache, backache and dizzy spells are robbing this woman of youth, beauty and health'.[50] Women were clearly accustomed to the hysterical, nervous female stereotypes being used as a vehicle to sell these products, with themselves as target customers.

Advertisements for aspirin and other over-the-counter analgesics followed suit, helping to construct the image of nervous, female pain. An example from 1922 depicts a woman slumped over a table, face buried in her hands, with a man standing over her. The advertisement is for Genaspirin: 'The safe way to relieve headaches.' Although the headache is the main ailment, it is accompanied by 'throbbing nerves'.[51] As late as 1959, an advert for Dispirin alludes to the despondent woman. A couple are seated at a dinner table and the man suggests 'will you have an aspirin ... I mean a Dispirin?'[52] In this example no specific, pain-related reason is given as to why the woman should require an analgesic. The image is loaded with the subtext of the despondent woman. There were also many claims that aspirin and related painkillers would work to alleviate insomnia, such as the aforementioned Genaspirin example, again making subtle reference to the hysterical nature of women's pain.

Compound analgesics and keeping going

The beginning of the twentieth century saw huge social change, in particular for women, from the decline of the corset to the right to vote. Meanwhile, hysteria and neurasthenia were losing their medical acclaim by the 1930s. However, the image of the hysterical woman did not disappear from advertisements for painkillers, which changed with the times in the mimetic fashion that Showalter describes. The idea of aspirin as a stimulant, as we saw in George Orwell's example, took on further importance in the twentieth century, although there was no clean-cut transition of aspirin from sedative to stimulant. Important in this shifting depiction of the woman in pain was the introduction of the compound analgesic in the early twentieth century. Compound analgesics typically feature one or more analgesics often combined with caffeine. An example which may be familiar is Anadin (known as Anacin in the USA). This drug, first available in the 1930s, originally contained aspirin, phenacetin and caffeine. The copy from a 1959 advertisement reads:

> 'Anadin' is like a doctor's prescription. It contains a combination of powerful active ingredients; aspirin, to relieve pain immediately, phenacetin to prolong the relief and calm jangled nerves, caffeine and quinine to combat depression and give a tonic effect. Anadin leaves you cheerful, relaxed and – of course free from pain.[53]

This example not only shows that Anadin promised to cure a multitude of symptoms that might have fallen into the categories of 'physical' and 'emotional' pain (here we see nerves and depression appear side by side), but also the promise of something else: the terms 'cheerful' and 'relaxed' allude to idea of happiness. These types of compound analgesics, known as APCs (aspirin, phenacetin and caffeine compounds) were popular in Britain, the USA and Australia in the 1950s and 1960s. Two famous brands available in Australia from the 1920s onwards were Bex and Vincent's, which were commonly consumed in powder form. Eileen Hennessey investigated the cultural factors contributing to the sharp rise of the consumption of these medicines in Queensland, Australia, in the 1950s, which led to kidney disease and death for thousands of women because of over-use. Her book *A Cup of Tea, a Bex and a Good Lie Down* takes its title from John McKellar's 1964 theatre revue.[54] The title entered the Australian vernacular, a phrase directed at 'neurotic' women who needed to calm down. Both products have been taken off the market, but it took many years for people to realise that it was the diuretic caffeine that was causing dehydration and finally kidney failure, combined with the hot climate of Queensland. Despite containing two analgesic compounds, aspirin and phenacetin, the drugs were not primarily consumed as painkillers for pain of a corporeal nature. Bex in particular was instead marketed specifically to women on the premise of happiness and vitality. One advertisement's key word was 'confidence'. When interviewing women (and the statistics show that Bex was overwhelmingly consumed by women), reasons for taking the drug were shown to be 'ill health, colds, headaches, frustrations, depression'. Hennessey states that 'clearly, these women were ingesting APCs as a stimulant, often in conjunction with other sources of caffeine such as Coca-Cola or coffee'.[55] The drugs themselves already contained caffeine, and their consumption in this manner created a perpetual cycle of dependency.

Hennessey discusses a variety of social and cultural factors that led to the popularity of APCs in Australia, which could be observed in much of the developed world after the Second World War. Following an increase in the number of working women during wartime, women were under pressure to resume traditional domestic roles and repopulate the country at the end of the war. All manner of household appliances, particularly the washing machine, were aggressively advertised as selling leisure time to women, when in fact they raised hygiene expectations. Women therefore devoted more time to household chores than they previously would have. To add to the pressure, much childcare literature of the 1950s placed a disproportionate emphasis on mothers being responsible for the development of their children.[56] Meanwhile, attitudes towards female nervous disorders seemed to have changed little since the turn of the century. Advertisements for APCs targeted women explicitly. One example cited by Hennessey is for 'Zans', another APC, which pictures a woman with a pot of tea and a packet of the medication. The advertising copy reads: 'Housework was such a drudge... but now – a cup of tea with 2 "ZANS" TABLETS and I feel ready to fly through the work!'[57]

Although, as noted before, the use of aspirin as a form of stimulant and as a kind of sedative for nervousness and insomnia may seem paradoxical, the two uses can be linked by another nineteenth-century pathological obsession: that of fatigue. The metaphor of *The Human Motor*, the title of Anson Rabinbach's fascinating book on modernity and fatigue, served as a model for a standard of health. Fatigue was a limit of the human-machine, but one that every effort was made to transgress. Hygiene studies surrounding the workplace became popular in the nineteenth century. Karl Marx was fascinated by the depreciation of the working human body, and in 1904 German physiologist Wilhelm Weichardt attempted to develop a human fatigue vaccine.[58] A body without fatigue was, and arguably still is, a utopian dream. However, fatigue was not limited to the corporeal. Neurasthenia covered all forms of nervous exhaustion: lack of energy, or insomnia caused by exhaustion, were common symptoms.[59] We could therefore read the dual use of aspirin and related drugs as stimulant and sedative as two responses to the same problem that characterised the nineteenth and twentieth centuries more broadly. Fatigue demanded a stimulant for one to keep going, as well as assistance in switching off. A St Joseph aspirin advertisement from 1956 caught the reader's attention with the line 'here's how to beat headachy housework fatigue'.[60]

It was not just APCs that were being marketed to women in this way in the 1950s and 1960s. Aspirin continued to be marketed for nerves, but the themes of keeping going, confidence and happiness also became apparent. The Australian brand of aspirin, Aspro, owned by Nicolas (who also owned Bex), was marketed in precisely this manner. One advertisement from 1957 states:

> Headache and pain need not rob you of a happy life. Go out when you feel like it! Enjoy your evenings and weekends without interference from nagging headache and pain. Look your loveliest at all times. Enjoy the admiration that comes from a happy, laughing expression unclouded by nerves and pain. You have to be well to be wanted. It's amazing what 'ASPRO', the genteel but powerful modern medicine, can do to keep you attractive. With 'ASPRO' there are no harmful after-effects, no 'let-down', 'ASPRO' brings swift relief from the dull nag of headache, a blessed relief that wipes away the disfiguring lines of pain.[61]

As well as promising pain relief, help with nerves and happiness, this brand goes one step further to sell the drug on the premise of attractiveness. It seems there were no limits to what Aspro could remedy. Although this advertisement may seem far-fetched, it captures a certain essence of our expectations of over-the-counter painkillers that is relevant today. We expect painkillers to go a step further than simply allowing us to get on with our everyday lives.

The pain barrier

From the advertisements examined between 1910 and 1959, we can see that the type of pain that aspirin and other over-the-counter painkillers claimed to target overlapped the categories of 'physical' and 'emotional'

pain with strong reference to pre-existing nineteenth-century pathologies of nervous disorders. These advertisements portrayed this type of
nervous pain as distinctly female. The brands featured women in their
campaigns, whilst also targeting their products towards women, sup-
porting the image of a nervous, hysterical woman that was constructed
in the late nineteenth century. She suffered from symptoms including various physical pains and insomnia. The absence of this type of
nervous pain from advertising targeted at men, and a distinct lack of
advertising aimed at the male consumer, reinforces this gendered construction of pain. The use of compound analgesics by women serves
as an illustration of the mimetic nature of hysteria described by Elaine
Showalter: as social conditions change, so do the characteristics of hysteria. In the case of APCs, hysteria, and thus a construction of women's
pain, was reshaped into an inability to cope with the new pressures of
post-war living.

In the trajectory of the advertising of over-the-counter analgesics in
this period, the body as machine can be seen to overlay the image of
the hysterical woman. Although we may think of the metaphor of 'the
human motor' as an image of the working, male body, the invisible
labour in the home was not exempt from the demands of industrialised
life. Resilience to fatigue was therefore central to female subjectivity with the increasing demands of work and leisure. From the most
recent examples of advertisements, it is clear that pain today must not
interfere with such demands: fast, effective pain relief is essential; we
cannot let pain form an obstacle to life. For example, the latest Nurofen
strapline is 'for lives bigger than pain'. The television commercial features men and women, all with interesting 'extreme' professions, for
example, a mountain pilot and a wildlife photographer.[62] The message
is that one can keep going not simply through the mundane everyday
tasks, but through demanding, boundary-pushing activities. This sense
of diminished tolerance to aches and pains culminates in McTavish's
observation of the headache as nuisance and David B. Morris' critique
of our 'immense faith in chemical assaults on pain'. How could these
two phenomena with regard to the consumption of over-the-counter
analgesics be more widely accounted for in the context of modernity?

Changing views of both health and pain can be seen to coincide with
broader historical shifts from the late nineteenth century. We have the
tendency to see pain and disease as invaders from the outside rather
than as a case of imbalance, as was the case in humoral theories of medicine. Alan Eherenberg argues that nineteenth-century neurasthenia
was central to this shift, owing to the theory of 'shocks', whereby the

impact of the outside technological world created a nervous individual. This, he argues, created the notion of the 'exogenous' whereby 'something that originates outside the individual creates a transformation within'.[63] If the aches and pains of modernity have their cause in the outside world, finding an internal cause loses its relevance, and anaesthesia seems a valid response to pain. Although we make a distinction between harder drugs such as opium or cocaine or even alcohol and the mundane aspirin, all these substances were consumed with the aim of counteracting the impact of modern life.[64] Therefore, we can also see aspirin, although its use was not recreational or hedonistic, as a drug to counteract these 'shocks' of modernity.

The second factor is linked to the idea of fatigue, a late nineteenth-century term that was initially described as 'the body's refusal to bend to the disciplines of modern industrial society'.[65] As work and fatigue became subjects of hygiene studies, health became defined as an ideal of the producer society. Zygmunt Bauman states:

Health is normative, drawing a boundary between 'norm' and 'abnormality'. Health is a desirable state of both the spirit and the body. To be healthy means to be 'employable': to be able to perform properly on the factory floor, to 'carry the load' with which work may routinely burden the employee's physical and psychical endurance.[66]

It is clear that the use of aspirin as a pick-me-up and stimulant would thus be appealing under such a conception of health. According to Bauman, health as an ideal for the industrial subject transformed into fitness, as modernity shifted from the 'solid' industrial times to the 'liquid' post-industrial times. Boundaries lost their importance. To be fit means to have a body that is 'flexible, absorptive, and adjustable'. Fitness is less definable than health as it remains 'permanently open', without limits. We can again see how advertising mirrors this ideal, from Aspro that promises limitless benefits to the contemporary Nurofen slogan 'for lives bigger than pain'. The ability to 'kill' pain removes a boundary and facilitates a limitless, utopian body.

From the 1960s onwards, we see almost a complete disappearance of emotional pain in the advertising of over-the-counter analgesics. Two reasons are briefly suggested here: first, the term 'depression' became a medically recognised condition, coined by Adolf Mayer and entering DSM-I in 1952 as 'depressive reaction', becoming 'depressive neurosis' in DSM-II in 1968. Since the 1950s, there has been a surge of antidepressant pharmaceutical products, and depression is defined as a chemical condition.[67] Thus, we distinguish between our painkillers: selective serotonin reuptake inhibitors (SSRIs) such as Prozac for depression, or sedatives such as Valium for anxiety disorders, and NSAIDs for pain of the corporeal nature. Second, in 1971, the 'scientific' explanation of how aspirin works was discovered by John Vane. Aspirin inhibits the production of prostaglandins, the chemicals released that enable us to feel pain.[68] It is perhaps following this discovery that we see the language of advertising become more 'scientific' and focused on physical symptoms. There are signs that this trend could again reverse. With new understandings of the involvement of affective centres of the brain in the experience of pain, over-the-counter painkillers are now being tested as emotional analgesics.[69]

Alongside the disappearance of the use of aspirin for emotional pain came the disappearance of the image of the neurotic or hysterical woman in the advertising of painkillers. However, Morris has argued that a culturally constructed, 'gender-marked pain' has instead gone 'underground'.[70] Indeed, if we are to follow Showalter and Morris, we cannot accept that the hysterical woman is absent from our understanding of pain today. Whether she resides in the discourse of chronic fatigue syndrome or of anorexia, or in a particular framing of female depression, it remains for scholars in the medical humanities to continue to analyse the contemporary entanglement of biological and cultural aspects of pain and disease.

Notes

1. 'Pharmacy Restaurant & Bar', www.damienhirst.com/projects/1998/pharmacy-restaurant (date accessed 24 February 2014).
2. Nurofen 200 g Liquid Capsules, purchased March 2012, manufactured by Reckitt Benckiser Healthcare (UK), Ltd, Slough, SL1 4AQ.
3. Diarmuid Jeffreys, *Aspirin: The Remarkable Story of a Wonder Drug* (London: Bloomsbury, 2005); Jan R. McTavish, *Pain and Profits: The History of the Headache and its Remedies in America* (New Brunswick: Rutgers University Press, 2004).
4. McTavish, *Pain and Profits*, 2.
5. This is described particularly in the case of cancer by Susan Sontag, *Illness as Metaphor* (New York: Farrar, Straus & Giroux, 1978), 3.
6. McTavish, *Pain and Profits*, 17.
7. David B. Morris, *The Culture of Pain* (Berkeley: University of California Press, 1991), 2.
8. *Ibid.*, 60.
9. *Ibid.*, 61.
10. *Ibid.*, 65.
11. *Ibid.*, 5.

12. Michel Foucault, *The Will to Knowledge: The History of Sexuality*, vol. 1 (London: Penguin, 1988).
13. *Ibid.*, 13.
14. *Ibid.*, 13, 105.
15. *Ibid.*, 30.
16. Jeffreys, *Aspirin*, 13–15.
17. McTavish, *Pain and Profits*, 69.
18. *Ibid.*, 70.
19. *Ibid.*, 72.
20. John Blake White, 'Antipyrin as an Analgesic in Headache', *Medical Record*, 30 (1886): 293, cited in McTavish, *Pain and Profits*, 80.
21. McTavish, *Pain and Profits*, 80.
22. Jeffreys, *Aspirin*, 73–7. There is discussion as to whether the invention is attributable to all three scientists or Dreser alone. Jeffreys argues that Dreser omitted Eichengrün and Hoffman from his pre-launch paper, 'Pharmacological Facts about Aspirin (Acetylsalicylic Acid)', but all three were involved in the synthesis of aspirin.
23. *Ibid.*, 126–7.
24. *Ibid.*, 177.
25. McTavish, *Pain and Profits*, 139.
26. Isabelle Leighton (ed.), *The Aspirin Age 1919–1941: The Essential Events in American Life in the Chaotic Years between the Two World Wars by Twenty-Two Outstanding Writers* (New York: Touchtone Books, 1949), 1.
27. Jeffreys, *Aspirin*, 124–208.
28. *Manchester Guardian*, 23 May 1919.
29. Susan Buck-Morss, 'Aesthetics and Anaesthetics: Walter Benjamin's Artwork Essay Reconsidered', *October*, 20 (1992): 3–41 at 19.
30. George Orwell, *The Road to Wigan Pier* (London: Penguin, 2001), 89.
31. Wolfgang Schivelbusch, *Tastes of Paradise: A Social History of Spices, Stimulants, and Intoxicants* (trans. David Jacobson) (New York: Vintage Books, 1992), 19.
32. Deborah Lupton, 'Food and Emotion', in Carolyn Korsmeyer (ed.), *The Taste Culture Reader: Experiencing Food and Drink* (Oxford: Berg, 2005), 319.
33. Jeffreys, *Aspirin*, 72.
34. James Harvey Young, *The Toadstool Millionaires: A Social History of Patent Medicines in America before Federal Regulation* (Princeton University Press, 1961), 9.
35. Jeffreys, *Aspirin*, 81.
36. *Ibid.*, 80.
37. *Ibid.*, 81.
38. Susan Strasser, 'Commodifying Lydia Pinkham: A Woman, A Medicine, and A Company in a Developing Consumer Culture (Working Paper #32)', 6, available at: www.consume.bbk.ac.uk/news/publications.html#workingpapers (date accessed 24 February 2014).
39. Peter Melville Logan, *Nerves and Narratives: A Cultural History of Hysteria in Nineteenth-Century British Prose* (Berkeley: University of California Press, 1997), 18.
40. *Ibid.*, 28.
41. Mark Pendergrast, *For God, Country and Coca-Cola: The Definitive History of the World's Most Popular Soft Drink* (New York: Thomson Texere, 2000), 9.

42. For more detail, see Ruth E. Taylor, 'Death of Neurasthenia and its Psychological Reincarnation', *British Journal of Psychiatry*, 179 (2001): 550–7.
43. Elaine Showalter, *Hystories: Hysterical Epidemics and Modern Culture* (New York: Columbia University Press, 1997), 15.
44. Foucault, *History of Sexuality*, vol. 1, 104.
45. Mark Micale, *Hysterical Men: The Hidden History of Male Nervous Illness* (Cambridge, MA: Harvard University Press, 2008), 123–53.
46. Showalter, *Hystories*, 33.
47. Elaine Showalter, 'Hysteria, Feminism and Gender', in Sander L. Gilman *et al.* (eds), *Hysteria Beyond Freud* (Berkeley: University of California Press, 1993), 286–344 at 317–19.
48. Morris, *Culture of Pain*, 104.
49. Pendergrast, *For God*, 9.
50. The advert was widespread across North America. See, for example, *Amarillo Globe-Times*, 17 March 1932.
51. *Manchester Guardian*, 14 March 1922.
52. *The Observer*, 9 July 1950.
53. *The Observer*, 24 May 1959.
54. Eileen Hennessey, *A Cup of Tea, a Bex and a Good Lie Down* (Townsville: James Cook University, 1993).
55. *Ibid.*, 77.
56. *Ibid.*, 54–70.
57. *Ibid.*, 70, emphasis in original.
58. Anson Rabinbach, *The Human Motor: Energy, Fatigue, and the Origins of Modernity* (New York: Basic Books, 1990), 142.
59. *Ibid.*, 153.
60. The ad appeared in *Star-News*, Wilmington, North Carolina, 10 December 1956.
61. Advert for ASPRO from *Australian Women's Weekly*, 13 March 1957, 59, cited in Hennessey, *Cup of Tea*, 82.
62. Nurofen, 'For Lives Bigger Than Pain', 11 February 2013, available at: www.youtube.com/watch?v=xyyzV2jD1sM (date accessed 24 February 2014).
63. Alan Eherenberg, *The Weariness of the Self: Diagnosing the History of Depression in the Contemporary Age* (trans. David Holmes *et al.*) (Montreal: McGill-Queen's University Press, 2010), 30.
64. Buck-Morss, 'Aesthetics and Anaesthetics'.
65. Rabinbach, *Human Motor*, 38.
66. Zygmunt Bauman, *Liquid Modernity* (Cambridge: Polity, 2000), 77.
67. See Eherenberg, *Weariness of the Self*, 147 for more detail.
68. Jeffreys, *Aspirin*, 230.
69. D. Randles, S.J. Heine and N. Santos, 'The Common Pain of Surrealism and Death: Acetaminophen Reduces Compensatory Affirmation Following Meaning Threats', *Psychological Science*, 24(6) (2013): 966–73; see also the press release 'Experiencing Existential Dread? Tylenol May Do the Trick', Association for Psychological Science, www.psychologicalscience.org/index.php/news/releases/experiencing-existential-dread-tylenol-may-do-the-trick.html (date accessed 24 February 2014).
70. Morris, *Culture of Pain*, 104.

9
Body, Mind and Madness: Pain in Animals in Nineteenth-Century Comparative Psychology

Liz Gray

Pain is a temporal entity that belongs neither in the history of the passions nor in the history of science, but rather 'between the world of emotions and the realm of sensations', in the history of experience.[1] The experiential nature of pain presents a challenge of subjectivity to both contemporary and historical study: how can we truly understand another's experience of pain? When questions of animal pain are addressed, the historian faces greater problems still. The complexity of interspecies subjectivity is compounded by anxieties of anthropomorphic validity. If we cannot fully understand the pain of another human being, who expresses emotional and/or sensational feelings through familiar words or contortions of the face and/or body, then how can it be possible to know what an animal is feeling? It is a state where neither the quality nor the quantity of the feeling can be understood. To begin to address this complexity, this chapter focuses on Scottish naturalist-physician William Lauder Lindsay's contribution to the theories of comparative psychology, setting out to locate his approach between the emotional and sensational worlds identified by Moscoso. It argues that animals' capacity subjectively to experience emotional or mental pain came to be accepted in part because of a prior acknowledgment that they suffered physical pain. The roots of comparative psychology are to be found, in part, within the history of physiology and its understanding of sensational pain.

Nineteenth-century comparative psychology was a diverse discipline, the term having been adopted by anthropologists, physiologists, mental scientists (now known as psychiatrists) and psychologists, with each group defining it differently. For some of those individuals, the subject included ideas of intellect and mind in the lower animals, with comparisons made with and between non-human or 'lower' species, as well as with and between humans (as 'higher' animals). This chapter places Lindsay's *Mind in the Lower Animals in Health and Disease* (1879) into the context of the shifting ground of research into the expression of emotions before and after the publication of Charles Darwin's *The Expression of the Emotions in Man and Animals* in 1872, the rise of physiological research from the 1870s, and other seminal works in early comparative psychology, such as Darwinian disciple George John Romanes' *Animal Intelligence* (1882). Historical work in this area has largely been focused on the naissance of Darwinian theory and the role of the animal in both the laboratory and society.[2] The recent emotional turn in history has not been neglectful of the issue of the animal mind.[3] The focus on the sympathetic, compassionate and humanitarian emotions of the vivisector in particular has brought the issue of animal pain into this historical discourse. This chapter aims further to expand this field from its physiological setting into the psychological.

Nineteenth-century physiologists who carried out vivisectional experiments did not deny that animals could experience pain. Physical responses to painful irritation, such as a frog wiping away a noxious liquid (one of the key experiments used in the development of the theory of reflex action), provided the physiologist with evidence for a discussion of the nervous system. In this case, pain and reactions to it were the focus of the study. By and large, however, the capacity of animals to experience pain was tacitly acknowledged in the usage of anaesthesia to remove physical pain in the animal. The anaesthetised animal became an object to be observed rather than a feeling subject. The physical pain of animals might have provided important insights into human physiology, or else it could be eliminated.

Comparative psychologists tended to accept physiological research and turned their attention to the subject of animal intelligence – a wide-ranging term that covered the reasoning, instinctual and emotional abilities of all animals 'below' 'man'. The experience of pain falls into the latter two of these three categories and is therefore the focus of this chapter.

Physical pain and instinct

The expression of pain in the animal as a subject of study appears in the first article on the disciplinary approach of comparative psychology published in Britain:

> The language of action which all animated beings are compelled to exhibit, being independent of their volition, and arising uniformly

the same under the pressure of similar circumstances, it is less susceptible of change by its own nature, than that which is purely con-
ventional [such as language]; and as it is understood, by a direct act
of consciousness, identifying in ourselves the connection between
the feelings and their natural expressions, there is no room for error;
the instant we can judge our own sensations, and of the changes in
our exterior which accompany them, it is impossible that we should
be mistaken in translating the expressions of others ... The writhing
of a worm when trodden upon is as clearly indicative of the pain it
experiences, as the gesticulations of the happiest actor; and the flight
of the hare or the roar of the lion are perfectly understood by every
living being interested in the intellectual movement of those animals.[4]

The anonymous author asserted that by understanding our own feelings and their related expression, we could, by analogy, understand the
feelings of the animal in our gaze by its expression. The writhing of
the worm was clearly, for this author, an expression of physical pain.
Published in 1820, this article explored a 'science' of movement and
expression that provided an insight into the animal mind. Following
this description of the expression of pain, the article took a physiological turn, exploring the mechanism by which the sensations of pain,
happiness, fear and anger were expressed.

The worm as experimental subject, and pain as the observable object,
was taken up 18 years later in the *Transmutation Notebook* of Charles
Darwin, but the insight into the animal mind was lost. Darwin related
the physical reactions to pain to inherited instinctive responses:

Even the worm when trod upon turneth, here probably there is no
feeling of passion, by muscular exertion consequent on the injury &
consequently excited action of heart – now this is the oldest inherited
and therefore remain, when the actual movement does not take place.[5]

Darwin concluded this observation with the statement: 'A start is
habitual movement to danger.' For him, the bodily reaction to potentially dangerous stimuli was an instinct towards survival, with physical
injury, or its threat, prompting a physical reaction.

Darwin returned to the subject of animal expression in 1872 with
the publication of *The Expression of the Emotions in Man and Animals*.
This text was, and still is, an important document to those studying the mind of animals. It was an answer to, but also a significant
departure from, Charles Bell's *Anatomy and Philosophy of Expression*.

Both texts were based on the science of observation and provided the basic descriptions of the objects of comparative psychological study. As Joe Cain notes in his introduction to the edition of *The Expression* republished in 2009: 'The focus on expressions allowed him [Darwin] to neatly bypass speculative questions about the emotional lives of animals.'[6] Accordingly, Darwin's allusions to pain were limited to instinctive responses to physical pain. Alongside the bodily writhing, groaning and the gnashing of teeth that appear as expressions of pain, physiological responses are also detailed: 'Perspiration bathes the body, and drops trickle down the face. The circulation and respiration are much affected.'[7] This physiological observation is accompanied by a survey of contemporaries' physiological understandings of the relationship between nerve activation and muscular movement, with a much clearer statement on the behaviour of animals when in pain. The strong desire to escape pain instilled a (heritable) habit of muscular exertion that acted most strongly on the muscles most often used. Since these included the muscles of the chest and the vocal cords, Darwin attributed the vocalisation of pain in animals only to a flight response, the advantage of serving as a call for help or a warning of danger being unintentional.[8]

Nineteenth-century physiologists and comparative anatomists were using this non-emotional conception of instinctive responses to pain in the development of their physical and physiological evolutionary scale from the anatomically simple to the anatomically complex, exemplified by 'man'. The (attempted) avoidance of discussions of an animal's emotional life whilst discussing their emotional expression is in stark contrast to the work of George John Romanes. Romanes, who thanked Darwin for providing him with 'all the notes and clippings on animal intelligence' that he had collected (positioning himself as the recipient of Darwin's legacy in terms of the animal mind), included several paragraphs on the emotionality of some (but not all) the animals he studied alongside his writings on their 'General Intelligence' in *Animal Intelligence*.[9] Romanes explicitly emphasised the emotions of certain animals, while folding the emotions of other animals into a more general discussion of their intelligence. He relied on a classical understanding of a hierarchy of species, which he based on mental and moral development – a scale that could be described as a 'psychological hierarchy'. At the higher end of the scale, certain animals were capable of grief and jealousy, affective states that had clear parallels with what would colloquially have been called 'pain' at the emotional level in humans.[10]

Pain and the mind

The charting of the rise of the 'emotions', in contrast to the 'passions' by Thomas Dixon in *From Passions to Emotions*, locates a shift in vocabulary towards one all-encompassing term in the nineteenth century.[11] 'Emotions' replaced a multifarious collection of terms that we (in the twenty-first century) might see as synonymous – such as affections, passions and feelings. The understanding of these terms by the authors of the comparative psychologies upon which this chapter is focused is never defined. In fact, the terms appear to be used interchangeably. Neither Romanes nor William Lauder Lindsay explored or explained the theory of emotions upon which they based their work; instead, they both appeared ready to accept the term, alongside passions, feelings, appetencies and sensations, as being known and understood by their readers. Dixon notes that Darwin did not have a theory of emotion either, but did, on occasion, incline to suggest that emotions were 'constituted by their expression'.[12] But whilst Darwin and Romanes did not differentiate between 'types' of emotion, for Lindsay this was an important part of his approach to comparative psychology. Lindsay's interest in the health of the animal mind stemmed from his experience as physician at the James Murray Royal Lunatic Asylum in Perth, Scotland. His approach to the diagnosis of his human patients focused as much on emotional pathologies as physical pathologies, and this was transferred to his study of the animal mind. In *Mind in the Lower Animals in Health and Disease*, the pathological emotions are divided into two main groups: those relating to pleasure and those relating to pain.[13] These differing types of emotion were expressed by different behaviours, and an excess of either sort could produce a change in behaviour that reflected a detrimental change to the mental health of the animal ('higher' or 'lower'). Negative emotions, such as grief, despair or loneliness, could have quite marked effects on behaviour and at relatively low levels of experience. They therefore posed a risk to any animal or human.

Whilst comparative psychology had moved along from the instinctual pain recognition of the 1820 article in the *New Monthly Magazine*, observation of an animal's behaviour remained the means by which the feelings of the animal could be understood, and a key tenet of the discipline. But physiological understandings of instinct were replaced by the need for physiological explanations of mind. By the time *Mind in the Lower Animals* was published, a renewed debate about the location and substance of the mind was well underway, and was the purview of physiologists, alienists/mentalists and psychologists. Men such as Thomas

Laycock, Henry Maudsley, William Carpenter, Herbert Spencer and Alexander Bain were all publishing explanations for mental phenomena, and Lindsay mined their work for theories to support his own.[14]

Lindsay was clear on his understanding of the concept of mind as it related to both man and the lower animals. This was a period when terms such as 'mind' (as already seen in emotion) were in a state for flux, and Lindsay initially left his readers with the option of using whatever definition or understanding they preferred for the term (again as with emotion). Yet later in the text he explains that mind had to be located throughout every human or non-human animal's body. This explanation of mind is important in order to understand Lindsay's conception of mental pain, and the physical expression by which it could be identified and observed. If mind was throughout the body, then influences on the mind, such as emotions, should be visible in the physical behaviour of the body. Parallels can be drawn here to the theory of emotions proposed by Bain and Spencer, described by Dixon as 'dual-aspect monism'. Underlying this theory was an understanding that there was a basic state connecting both the physical and the mental; in other words, 'that mental feelings and physical nervous processes were two sides of the same (unknowable) coin. The event could be looked at from the mental point of view or the physical point of view'.[15] Dual-aspect monism provides a theoretical framework with striking similarities to the ideas upon which Lindsay based his comparative psychology.

To return to the frog in the physiological laboratory wiping away a noxious substance, Lindsay's conception of the mind left him unconvinced by the notion that automatic responses to stimuli could completely explain the movements of acephalous animals. He believed physiologists were 'confounding ... mere reflex or *automatic* action with expressions of pain – for instance, in the decapitated frog'.[16] They denied the true experience of pain in the animal, when alternatively the corporeal awareness of physical pain could as easily explain the movements of brainless animals. If a creature had a corporeal mind, even if it had lost possession of its brain, then it was to be assumed that it had the ability to feel pain, and that this pain would be expressed through its behaviour.

As has already been suggested, Lindsay's comparative psychology was rooted in his experience as an asylum physician, the application of the understanding he gathered from observing his patients, his own observations of animals and the anecdotes of animal behaviour he had collected. In his published work on the treatment – both preventative and curative – of the insane, he promoted and followed the moral therapeutic approach. Where the physical restraint of human patients had been rejected for being damaging to both their physical and mental health, so the causing of physical pain to animals could also lead to the development of mental health problems. Any disease or injury could lead to temporary, and perhaps even chronic, mental changes. And Lindsay was not the only animal observer to draw this conclusion. Jean-Charles Houzeau, a French-born naturalist and astronomer who published a large volume on the mental faculties of animals, suggested that: 'All kinds of bodily suffering or *pain* may produce a kind of walking temporary delirium, which is but a stage towards mania.'[17] Lindsay expanded this point, with his typical reliance on anecdote:

> We read of the rhinoceros, elephant, reindeer, and other large animals being 'maddened by pain', be it toothache or fracture, the bites or stings of insects, or the wounds inflicted by man. The expression, so frequently used in reference to our domestic or menagerie animals, of being 'mad' with pain probably describes a furiosity that is apt to pass into, if it be not sometimes a transient, *mania*.

The movement away from an external source of pain could mask an emotional reaction. However, when the pain was internal, the physical response in terms of behaviour mirrored, as Lindsay saw it, the behaviour that resulted from insanity – the behaviour he saw in some of his patients.

A language of pain

If there were a link between the mental and the physical that was 'unknowable', if mind had the ability to express itself through the body, then it would be possible to discuss this expression/behaviour in terms of a language of pain.

In the 1820 *New Monthly Magazine* article, a 'universal language of nature' was argued to be the foundation of the methodology upon which the science of comparative psychology could be based.[18] The language was one based on expression and one that could cross the species barrier. This outline presented a science that would depend on the skills of observation and factual interpretation, and as such was one that suited the naturalist as much as it suited the anatomist and physiologist.

In his *Anatomy and Philosophy of Expression as Connected with the Fine Arts* (1806), Charles Bell drew a difference between humans and animals in their abilities of expression. Bell acknowledged that whilst animals were able facially to express the most basic passions and emotions, once these emotions became more complicated – that is, once they became 'character' – they became the purview of humans. His anatomical dissections of both humans and animals demonstrated to him that animals lacked some musculature that prevented them from expressing these characters. When it came to the expression of pain in particular, where the animal would retract its lips and expose its teeth, man's facial musculature would half close the mouth. This expression was one representative 'more of agony of mind than of mere bodily suffering, by a combination of muscular actions of which animals are incapable'.[19] There was a fine line between passion and character, but in terms of mental pain, the animals anatomically could not cross it. Physical pain was bodily, whereas other expressions of pain, such as weeping, were mental and relied on specific musculature; they altered the shape of the face in characteristic ways. In this way all animals, including man, could express pain, yet weeping was a 'character' of human pain. Weeping was mental rather than physical.

Bell wrote in greater detail about the relationship between pain and fear, and the often-conjoined nature of their expression. In these cases, the fear is fear of pain. All animals could feel this passion: the dog feared the pain of a beating just as man feared being burnt or injured. The difference in expression was not to be found in its countenance, but in its bodily expression. Pain would energise and create a level of tension in the body, whereas fear would paralyse the muscles. Fear and bodily pain were rooted in man's animal nature; they were basic expressions. Mental pain and terror, which Bell explained as being 'peculiar to man', both required the mind to produce them. Many of the negative emotions, such as terror, despair and jealousy, were presented as being unique to man – requiring both a mind and specific anatomical structures in order to experience them:

> Of man alone that we can with strict propriety say, the countenance is an index of the mind, having expression corresponding with each emotion of the soul. Other animals have no expression but that which arises by mere accident.[20]

Both Darwin and, before him, William A.F. Browne attacked Bell's thesis of this 'difference in kind' between man and animals.[21] The worldview of anatomy, physiology and psychology had undergone dramatic changes in the decades between the first publication of Bell's work and Darwin's *Expression of the Emotions*, in most part because of Darwin himself. There was a growth in the acceptance that any difference between man and the lower animals was one of degree, not of kind. Animals had minds and, as Darwin was aiming to prove, some of them had anatomies and physiologies that resembled those of man. As already noted, Darwin attempted to avoid speculation about the meaning or experience of emotions from his work and focused solely on the physiology behind emotional expression. As Dixon notes, the communication that was ascribed to the expressive movements of animals was, for Darwin, a mere byproduct of behaviours that originally had no communicative purpose.[22] In the earlier discussion of his work on bodily pain, physical movement was indicative of this sensation – the writhing of the body, the gnashing of the teeth. Vocalisation, in the form of groaning, was also included. Vocalisation of these sorts were also included in Bell's work, but in the descriptions of fear rather than pain.[23] Darwin differentiated between the vocalisation of the animal and the language of man. Whilst noting the 'force of language is much aided by the expressive movements of the face and body', he referred in his concluding remarks to the 'articulate' language of man as different from the vocalisation of sound by animals.[24] Others, however, perhaps influenced as much by anthropologists as by physiologists, anthropomorphised the expressions they were observing and did not limit language to man.

Lindsay's introduction to the interpretation of 'animal language' came from a lecture given by Professor John Goodsir, Edinburgh Chair of Anatomy, in 1856 entitled 'On Life and Organisation'. Within this, and a discussion of the new psychological science, he laid out the mechanism by which an animal's 'feelings' were conveyed:

> The various signs and noises indicative of the Appetites, Affections, and Passions of the lower animals constitute, indeed, an elementary form of Language; but it is entirely destitute of the discursive element which, distinctive of thought, exhibits itself in the relative terms of logical speech. The so-called Language of the brute is merely a succession of signs, each sign significant of a particular appetite or emotion, and primarily induced therefore by an objective excitement.[25]

The importance of this statement is the acknowledgment that the language of animals takes two forms: the audible and the physical.

Lindsay made use of the audible aspects of this 'language of the brute', dedicating a chapter to the subject within his book: 'Pain is frequently expressed in other animals, as in man, by *yelling. Moaning* is an equally common expression of physical suffering and mental grief.'[26] Whilst this could be read as being similar to the vocalisations of pain discussed by Darwin, Lindsay ascribed a clear communicative function to the sounds and, as such, departed from Darwinian theory (which was, in the *Expression*, in its most Lamarckian mode). However, in referencing the work of Houzeau, he expands and defines this language further, noting that different animal species possessed different methods of expressing different emotions. The terms 'yelling' and 'moaning' only represented a simplistic overview: 'The *howl* of a dog may proceed from (a) bodily pain; (b) loss of way or master; (c) any kind of disappointment; (d) anger, grief, despair, or even mere impatience.'[27] As will be addressed shortly, (b), (c) and (d) could all be interpreted as examples of mental pain. Lindsay continued: '*Wailing* may arise from bodily pain, grief and expostulation.' These vocalisations of the dog were then compared with those made by the cat: 'The mewing of the cat may express anguish, sadness, and melancholy – the result of ungratified love ... *Moaning* may be a sign equally of grief, of mental pain, or of that which is purely physical.'

According to Lindsay, the belief that humans were the only species to possess language had resulted in the audible language of animals being less well studied than other forms of communication. Therefore, it was of less use in the study of emotions in the lower animals than it ought to have been. In what appears to be typical within Lindsay's reasoning, he ignored the problem that the vocalisation of physical and mental pain could be easily confused. The main focus of his text was on mental pain, and therefore he discounted the possibility that physical pain could be the sole cause of these sounds. It is worth speculating that Lindsay thought physical pain caused mental pain in these instances. For if the mind was expressed throughout the corporeal, then bodily injury would in turn cause 'pain' of the mind, and as such be expressed in the same way.

Lindsay called for the increased teaching and study of comparative language, and not just comparative philology, but other 'forms' of language as well. The 'brute' language, as he described it, included both vocal and physical elements and varied between species. For example, the language of the dog consisted of four separate attributes: the voice (as discussed above); the eye and the look; the tail and ear; and the general attitude of the animal. The more useful language had a physiognomic basis – it was visual: 'The dog's eye is not less expressive of disease or pain, mental or bodily, than of pleasurable emotions.' Whilst physiognomy was becoming outdated in disease diagnosis, it was a vital aspect of animal study and for some alienists it still had a role to play in mental disease.

Towards mental pain

The moral hierarchy of society that had been modified in the wake of Darwinian theory was used by the comparative psychologist to rank animals against humans according to their perceived mental standing.[28] Some animals were seen as being above 'savage' humans – domestication and breeding changed the behaviour and therefore the moral standing of these animals.[29] Romanes used the response to pain as an example of how these standings could be determined. The fox or the wolf (classed in this instance as 'wild' dogs) was seen to be able to suffer pain in silence, in contrast to the domestic dog being so susceptible to pain that it would 'scream' when suffering the mildest discomfort.[30] Here sensitivity was seen as being of greater moral value than what was perceived as insensitivity. Lindsay saw that this increased sensitivity to physical pain was mirrored by a greater susceptibility to the suffering of mental pain.

Nineteenth-century comparative psychologists used the term 'mental pain' as an overarching term for the many painful emotions that were suffered by man. Grief, loss, fear, anxiety, anger, envy, jealousy, hatred and despair are just some of the feelings Lindsay listed as having negative effects on the mind. They were described as moral causes of mental disorder and were therefore treated with the moral treatment approach common in asylums in that century. The history of madness has charted the search for the pathological, the social and the moral causes for mental disorders and derangements.[31] The early forms of comparative psychology discussed in this chapter, and in particular the work of Lindsay, is embedded in the moral approach where, despite searching, no one pathological cause for the pain could be identified and therefore no physical (as in surgical or medical) treatment was available.

As a physician, Lindsay oversaw the transition of his asylum from one that cared for patients from a variety of social classes to a private institution. His early years at the asylum provided him with opportunities to compare the moral behaviour of his rich and poor patients. His work on the animal mind focused, mainly, on the domestic animal, which he saw as akin to his richer patients. These more sensitive animals reacted to mental pain in a similar way to physical pain:

> Some dogs are in misery so long as they feel or see themselves under a master's displeasure. In other words, just as pleasure is the result of man's approval, pain is the effect of his disapproval. Man's blame may be conveyed in specially offensive and irritating ways, in taunts or upbraidings, associated with ridicule or sarcasm, in which case the moral result is much more serious than in ordinary reproof. Sarcastic

taunts are only too apt to sink deeply into, and to rankle in, the minds of sensitive animals.[32]

He took this further in the following paragraph: '*moral* proof or *punishment* is frequently felt to be more severe than that which is simply or directly physical'. These quotations appear in a chapter entitled 'Sensitiveness', a chapter full of examples of animals suffering mentally as a result of neglect, discouragement, humiliation and disgrace. Lindsay was not unique in this point of view; nor was he the only alienist. He referenced Dr Herbert Major, from the West Riding Asylum, who observed similar behaviour in his pet dogs when they were reprimanded.

Of the possible reactions to such mental pain outlined by Lindsay, a feeling of misery was a mild response. In more severely felt cases it could result in loss of affection for their master, the desertion of home and, most serious of all, fatal convulsions, or sudden or prolonged death. The majority of examples of death resulting from mental pain resulted from fear, but heartbreak from loss or desertion was also observed in a range of animals and noted as their causes of death. Lindsay declined, quite openly, to provide or even explore the physiological basis of these physical results of emotional causes. For him, this was an area of investigation for the physiologist, not the comparative psychologist, for the role of the latter was simply the elucidation of the connection between the 'mental and moral state and bodily effect'. And this was to be done through the interpretation of animal behaviour.

Although the suggestion of death as a direct result of emotional pain is the most dramatic pathological response, it was not for Lindsay the most interesting. The real interest lay closer to the symptoms he saw in his asylum patients. The motor disorders, such as epilepsy, were seen to make their first appearances within the muscular system, although it was understood that this was the result of the influence of, or through, the nervous system. The root causes in his understanding of these symptoms were the emotions. As a result of his understanding of the mind, there was no separation of physical and psychological pain. They were both interlinked sensations and both played a role within comparative psychological study.

Pain as an object of scientific study

For comparative psychologists such as William Lauder Lindsay and George John Romanes, the emotional capabilities of animals were indicators of their intelligence. Emotions were seen as part of the moral character of each animal or, more accurately, each animal species, which could be observed alongside behaviour indicative of abilities to reason and to learn. Pain could be both physical and emotional, and its physical nature meant that it could be suffered by all non-vegetative organisms. For comparative studies, therefore, it offered itself as a good object of study. The observation of physical pain crossed disciplinary boundaries, appearing as much in the results of physiology as it did in these comparative psychologies. Helen Blackman's study of the development of late-nineteenth century animal morphology outlines the case for the contingent growth of morphological and zoological sciences alongside the experimental laboratory sciences.[33] The writing of the histories of the biological and human sciences has focused on the processes of disciplinary formation and professionalisation, and therefore the decline and removal of the amateur from the ranks of the 'gentlemen scientist'.[34] Those undertaking observational and 'museum-based' sciences were gradually sidelined, bringing to the fore the more technical and equipment-based sciences. Blackman writes that 'once biologists' rhetoric was unpacked, morphology and physiology were not always clearly separated'. This chapter demonstrates that in relation to the psychological sciences, observation and the use of expressions of emotion as 'scientific data', this sentiment is particularly accurate.

The identification of negative emotions, and the interpretation of them as painful, enabled them to be understood as diagnostic tools within the asylum setting. The defining of mental pain was important to the moral therapeutic regimes of the asylum. Mental pain could be the cause of both physical and mental pathologies, and the moral treatments had to address these. For Lindsay, the developments in evolutionary theory had obvious correlations to the moral underpinnings of the understanding of mental disease. With the *Origin of the Species, The Descent of Man* and *The Expression of the Emotions*, Darwin had linked the 'higher' (humans) with the 'lower' animals. The social importance of sympathy, as discussed by Boddice and White, moved the issue of suffering and, in particular, pain into the experimental sciences of the day.[35] As research into the functioning of the nervous system began to produce results, the issues of the presence, function and location of the mind were brought to the fore by some.

Animal psychology, as a discipline separate from natural history, was still in the relatively early stages of development, and those with an interest in it brought differing methodologies for its exploration. The author of the *New Monthly Magazine* article and Romanes provided their experiences of the physiological laboratory; William Lauder Lindsay drew on the management of the mental and physical health of the insane. For Romanes, the understanding of the animal mind was simply part of a journey towards a greater understanding of the human mind. It was the natural continuation of the comparative studies of physiology and anatomy. Lindsay argued alternatively that in a Darwinian society where sympathy was prized above other 'emotions', the effect of man's treatment of animals on their mental states as well as their physical states could be read in their behaviour.[36] Lindsay thought that this sympathy ought to have been a powerful driver in the treatment of animals within the physiological laboratory and in the treatment of all natural beings below the highest rank of men in all of society. The asylum, as a house for the mentally damaged of society, ought to have been a site for compassion and humanitarian care, as with every laboratory and even every home.

Notes

1. Javier Moscoso, *Pain: A Cultural History* (Basingstoke: Palgrave Macmillan, 2012), 2.
2. Rob Boddice, 'Vivisecting Major: A Victorian Gentleman Scientist Defends Animal Experimentation, 1876–1885', *Isis*, 102 (2011): 215–37; Lorraine Daston and Gregg Mitman (eds), *Thinking with Animals: New Perspectives on Anthropomorphism* (New York: Columbia University Press, 2006); Richard D. French, *Antivivisection and Medical Science in Victorian Society* (Princeton University Press, 1975); Jed Mayer, 'Ways of Reading Animals in Victorian Literature, Culture and Science', *Literature Compass*, 7(5) (2010): 347–57; Robert J. Richards, *Darwin and the Emergence of Evolutionary Theories of Mind and Behavior* (University of Chicago Press, 1987); Harriet Ritvo, *The Animal Estate: The English and Other Creatures in the Victorian Age* (Cambridge, MA: Harvard University Press, 1987).
3. Rob Boddice, 'Species of Compassion: Aesthetics, Anaesthetics, and Pain in the Physiological Laboratory', *19: Interdisciplinary Studies in the Long Nineteenth Century*, 15 (2012); Rob Boddice, 'The Historical Animal Mind: "Sagacity" in Nineteenth-Century Britain', in Robert W. Mitchell and Julie Smith (eds), *Experiencing Animals: Encounters between Animal and Human Minds* (New York: Columbia University Press, 2012); Jed Mayer, 'The Expression of the Emotions in Man and Laboratory Animals', *Victorian Studies*, 50(3) (2008): 399–417; Paul White, 'Sympathy under the Knife: Experimentation and Emotion in Late Victorian Medicine', in Fay Bound Alberti (ed.), *Medicine, Emotion and Disease, 1750–1950* (Basingstoke: Palgrave Macmillan, 2006).
4. Anon., 'Comparative Psychology', *New Monthly Magazine and Universal Register*, 14 (1820), 297.
5. Charles Darwin, 'Notebook M: Metaphysics on Morals and Expression', 1838, http://darwin-online.org.uk/content/frameset?itemID=CUL-DAR125.-& viewtype=text&pageseq=1 (date accessed 24 February 2014), emphasis in original in second quotation.

6. Charles Darwin, *The Expression of the Emotions in Man and Animals*, 2nd edn (London: Penguin Classics, 2009 [1890]), xxx.
7. *Ibid.*, 73. The physiological discussion of pain can be found at 72–6.
8. *Ibid.*, 74–5.
9. George John Romanes, *Animal Intelligence* (London: Kegan Paul, Trench & Co, 1882), xi.
10. In *Animal Intelligence*, species at the lower end were insects. The higher end of the scale began with fish and continued through many domestic and large wild mammals.
11. Thomas Dixon, *From Passions to Emotions: The Creation of a Secular Psychological Category* (Cambridge University Press, 2003).
12. *Ibid.*, 166.
13. W. Lauder Lindsay, *Mind in the Lower Animals in Health and Disease*, 2 vols (London: C.K. Paul, 1879).
14. It should be noted that whilst the materiality of the mind, let alone its location, was the subject of much debate, Lindsay's conception was less common. Although he drew on a range of contemporary work, he was selective in the elements he chose in order to present evidence for his theory. Relevant works by the listed authors include: Thomas Laycock, *Mind and Brain: or, The Correlations of Consciousness and Organization* (Edinburgh: Sutherland & Knox, 1860); Henry Maudsley, *Body and Mind: and Inquiry into their Connection and Mutual Influence* (London: Macmillan, 1870); William B. Carpenter, *Principles of Mental Physiology*, 3rd edn (London: Henry S. King, 1875); Herbert Spencer, *Principles of Psychology*, 2nd edn (London: Williams & Norgate, 1870); Alexander Bain, *Mental and Moral Science*, 3rd edn (London: Longmans, Green & Co., 1872). See Roger Smith, *Between Mind and Nature: A History of Psychology* (London: Reaktion, 2013) for a historical overview and further secondary reading on this debate.
15. Dixon, *Passions to Emotions*, 142–3.
16. Lindsay, *Mind in the Lower Animals*, vol. 1, 4. Also covered in vol. 2, 3–8, emphasis in original.
17. Lindsay, *Mind in the Lower Animals*, vol. 2, 207, emphasis in original. J-C. Houzeau published *Etudes sur les facultés mentales des animaux comparées à celles de l'homme* in 1872.
18. Anon., 'Comparative Psychology', 298.
19. C. Bell, *The Anatomy and Philosophy of Expression: As Connected with the Fine Arts*, 6th edn (London: Henry G. Bohn, 1872), 139.
20. Sir Charles Bell, *Essays on the Anatomy of Expression in Painting* (Longman, Hurst, Rees, and Orme, 1806), 88. By 1872, this text had been retitled *The Anatomy and Philosophy of Expression* and was in its sixth edition. This quote can also been refocused away from the soul (136–7): 'But besides the muscles analogous to those of brutes, others are introduced into the human face, which indicate emotions and sympathies of which the lower animals are not susceptible; and as they are peculiar to man, they may be considered as the index of mental energy, in opposition to mere animal expression.'
21. Gregory Radick, *Edinburgh and Darwin's Expression of the Emotions*, Dialogues with Darwin 13 (University of Edinburgh, Institute for Advanced Studies in the Humanities, 2009).
22. Dixon, *Passions to Emotions*, 166.

23. It is worth noting that Bell was writing for artists rather than scientists, and the absence of sound in sculpture or painting may explain the relative lack of discussion on vocalisation within his work.
24. Darwin, *Expression*, 325.
25. John Goodsir, *The Anatomical Memoirs of John Goodsir* (William Turner and Henry Lonsdale (eds) (Edinburgh: Adam and Charles Black, 1868), emphasis in original.
26. Lindsay, *Mind in the Lower Animals*, vol. 2, 178, emphasis in original.
27. Lindsay, *Mind in the Lower Animals*, vol. 1, 294, emphasis in original.
28. Charles Darwin, *On the Origin of the Species by Means of Natural Selection, or, The Preservation of Favoured Races in the Struggle for Life* (London: John Murray, 1859); Charles Darwin, *The Descent of Man, and Selection in Relation to Sex* (London: John Murray, 1871). For the transition between pre- and post-Darwinian conceptions of the Great Chain of Being, see Rod Preece, *Brute Souls, Happy Beasts, and Evolution: The Historical Status of Animals* (Vancouver: UBC Press, 2005).
29. Boddice, 'Historical Animal Mind', esp. 75–6.
30. Romanes, *Animal Intelligence*, 441.
31. Examples of texts covering these subjects include: G.E. Berrios and Roy Porter (eds), *A History of Clinical Psychiatry: The Origin and History of Psychiatric Disorders* (London: Athlone, 1995); Andrew Scull (ed.), *Madhouses, Mad-Doctors, and Mad-Men: A Social History of Psychiatry in the Victorian Era* (Philadelphia: University of Pennsylvania Press, 1981); Andrew T. Scull, *The Most Solitary of Afflictions: Madness and Society in Britain, 1700–1900* (New Haven: Yale University Press, 1993).
32. Lindsay, *Mind in the Lower Animals*, vol. 1, 297.
33. Helen J. Blackman, 'The Natural Sciences and the Development of Animal Morphology in Late-Victorian Cambridge', *Journal of the History of Biology*, 40(1) (2007): 71–108.
34. Examples of this can be found not only in Blackman but also in Ruth Barton, '"Men of Science": Language, Identity and Professionalization in the Mid-Victorian Scientific Community', *History of Science*, 41 (2003): 73–119; Adrian Desmond, 'Redefining the X Axis: "Professionals," "Amateurs" and the Making of Mid-Victorian Biology – A Progress Report', *Journal of the History of Biology*, 34 (2001): 3–50; N. Jardine, J.A. Secord and E.C. Spary (eds), *The Cultures of Natural History* (Cambridge University Press, 1995).
35. See note 3.
36. Lindsay, *Mind in the Lower Animals*, vol. 2, 313–70.

10
Down in the Mouth: Faces of Pain
Danny Rees

Communicating pain

The inspiration for the title of this chapter originates in James Gillray's 1804 print of a patient being bled by his doctor, exemplifying the communicative function of facial expression and its recognition in others (Figure 10.1). Conforming to Humoral Theory, based on the Ancient Greek writings of Hippocrates, bleeding was believed to help restore the body's natural, healthy balance of vital fluids. Gillray's draws a distinct contrast between the two faces, especially in their mouths. The patient, who cannot bring himself to look at the procedure, makes the perfect downward grimace of unhappy resignation. His eyes are also cast downwards. But equally well realised is the tension in the mouth of the doctor, the tight 'pursing' around the lips constricting the blood supply to make the area pale. Note how the doctor's eyes are wide open and his brows are raised, indicating his full attention and some trepidation.

The facial expressions are exaggerated and somewhat stylised, and the setting is reminiscent of the theatre in its positioning of the characters. What is difficult to gauge is the purpose of this print. Are the juxtaposed expressions meant to be a source of humour? The scene will have had a different meaning in 1804 and will have been appreciated in ways alien to a modern audience. It has become a piece of art history rather than a satirical comment on contemporary treatments of illness.

Modern 'readings' or interpretations of these expressions must necessarily take account of the context of the depiction of domestic bloodletting in Gillray's picture. If, however, the faces were to be viewed in isolation, the possible scenarios and causes of vexation become open-ended and this is important because the perceived cause of the distress influences

164

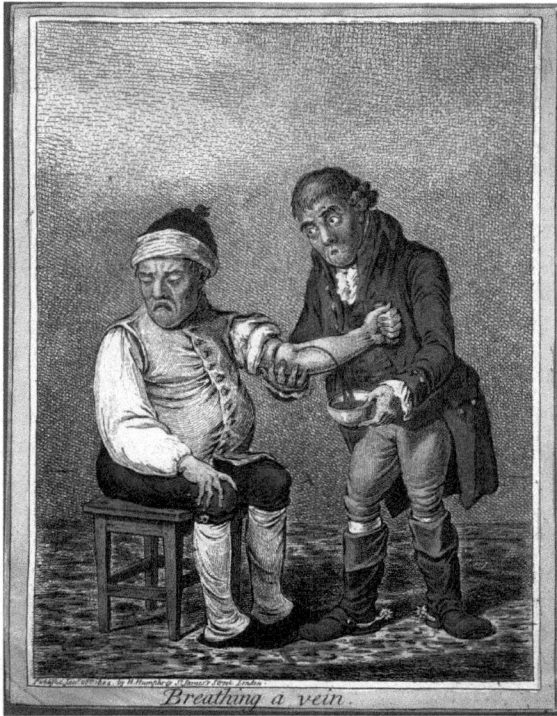

Figure 10.1 James Gillray, *Breathing a Vein*, 1804 (Wellcome Library, London)

the language used to describe these states. Thus, abdominal pain and grief may look similar, superficially, as facial expressions, but will not be described in the same terms and different values will be attached to these conditions. Indeed, grief itself may be represented differently in different times and places. In other words, without context, it is difficult to derive meaning from, or form judgements of, facial expressions.

As Rob Boddice points out, 'if we set out to look for examples of contemporary emotions in the past, we are certain only to find anachronisms'.[1] Referring to William Reddy's concept of the 'emotive', defined as an 'affective utterance' representing an individual's attempt to translate inward feelings through cultural conventions, Boddice suggests that these attempts to some extent are bound to 'fail' since inward feelings can never be 'authentically' expressed. Yet many of the examples used in this chapter treat facial expressions simplistically, as visual dimensions of emotions and as valid representations of internal states. They might conform to C.E. Izard's appraisal that: 'Emotion at one level of analysis *is* neuromuscular activity on the face.'[2]

In 2009, psychologist Kenneth Prkachin summarised the difficulties of studying facial expressions: they are 'complex, evolve over time and are difficult to describe and quantify. Normally they leave no record'.[3] Yet, despite their elusive nature, artists and scientists from a range of periods and perspectives have sought to classify and interpret different expressions and their meanings. Many of these investigations have attempted to 'fix' definitions of different emotional states and clearly delineate them. Part of those investigations has involved the subject of pain, both 'physical' and 'emotional'.

My focus is to highlight how examples of early anatomical and scientific experiments have tried to describe and define the face of pain, and how some of these influences have contributed to the concept of 'universality' as applied to emotional values supposedly represented as facial expressions. But why has the face been the locus of repeated, probing attention?

A functional account of expression would maintain that being able to 'read' the emotions of others is considered a vital route to successful social interactions, such as recognising when someone is distressed. Effective communication of that distress will aid the survival of those in need of help. The face is a key site of these visual processes and is also considered to be essential for relations among non-human animals: 'primate facial displays are evolutionarily designed devices to elicit a response from the receiver'.[4]

Acute pain or grief may produce crying and distort the features to provoke a sympathetic response in others, but the effect, for example, of crying in babies will elicit very different responses if you are sharing a train carriage rather than a close familial relationship with the infant in question. This is how Amanda C. Williams emphasises the function of social interaction regarding pain in her evolutionary account from 2002: 'the function of pain is to demand attention and prioritise escape, recovery and healing; where others can help achieve these goals, effective communication of pain is required'.[5]

Williams, a consultant clinical psychologist, posits that facial expression is one of the key non-verbal indicators of pain and claims that expressions remain consistent and recognisable whether in infants or in old age. Yet differences in cultural attitudes to expressing distress and societal changes over time are not the focus of her accounts. This chapter pursues the historical inconsistencies of approaches to expression and looks at the contingencies and assumptions required to authenticate or 'recognise' 'non-verbal indicators of pain'.

Painting pain

Representing an emotional state such as pain in art relies not only on the artist's skill but more significantly upon their appreciation of the subtle changes that alter the features. An early influential work attempting to denote a facial language of emotional expression was created by an artist greatly favoured in the French court of Louis XIV, Charles Le Brun. Thought to have taken place around the 1670s, he gave lectures on how to portray and categorise the 'appropriate' facial expression of different emotions. The artist's priority was to faithfully 'mimic' what could be observed in human actions and appearances. Le Brun looked to the Ancient Greek sculpture 'Laocoon and his sons' and the reaction of the family group as they are attacked by large snakes. From this sculpture, he used one of the figures to exemplify the term 'Acute pain' (Figure 10.2). It is understandable that he would refer to a moment 'frozen in time' and being captured in marble, as this allowed him

Figure 10.2   Charles Le Brun, *Acute Pain* (Wellcome Library, London)

unlimited time to study what would ordinarily be fleeting. Although he presented these expressions as single portraits out of context, he interpreted the results based on the original situation. Interestingly, a copy after Le Brun held in the Wellcome Library has the added handwritten note 'Symptoms of cholera', highlighting that such expressions can be thought a suitable illustration for an entirely different circumstance and cause.

In a later English translation of Le Brun's lecture, he notes that 'if sadness be caused by a sense of bodily pain, and that pain be acute all the motions of the Face will appear acute ... All the parts of the face will be more or less distorted, in proportion to the violence of the pain'.[6] Le Brun's representations of sadness depict the downturn of the corners of the mouth and the downward slope of the eyebrows from the middle of the forehead towards the ears. Superficially it resembles the expression of pain, although the mouth is distinctly less active and is not being pulled into a grimace.

Facial expression is presented by Le Brun as an external 'barometer' of the internal emotional 'weather', a gauge by which severity can be ascertained. From Le Brun's perspective, emotions are natural phenomena and the 'soul' unites all parts of the body to express the different 'passions', citing fear as expressed by a man 'flying away'. But if there is a part of the body that best shows this, it is the face and, in particular, the eyebrows:

> But if there be a part, where the Soul more immediately exercises her functions, and if it be the Part mentioned, in the middle of the brain, we may conclude that the Face is the Part of the Body where the passions more particularly discover themselves. And as the gland, in the middle of the brain, is the place where the soul receives the images of the Passions; so the eye-brow is the only part of the whole face, where the Passions best make themselves known; tho' many will have it to be the eyes.[7]

This reveals the influence of Descartes, who theorised that the pineal gland (the gland referred to by Le Brun) was the seat of the soul.[8] The proximity of the face to the brain may have helped determine its place as a 'window' to the interior world. But one cannot ignore the prevailing philosophies and religious beliefs that underpin how Le Brun explores the notion of 'passions'.

While Le Brun relied on observations of the outer appearance of expressions for his categorisations, the next major contribution to the field used the privileged surgeon's position of being able to go beneath the skin, performing dissections to reveal the structures underneath. Sir Charles Bell, surgeon, neurologist and anatomist, had closely examined the muscles of the face to further his medical studies, but offered his findings to artists in order that they might produce convincing portrayals of different emotional states. This work demonstrated the organisation and interplay of the muscles as seen and recorded by Bell, who was also responsible for the illustrations. Entitled *Essays on the Anatomy of Expression in Painting* and first published in 1806, it consists of a series of six essays over which he discusses the changing nature of the face from youth to old age, the shape and construction of the skull and facial musculature, before focusing on particular expressions in chapter five. Although basing his representations on knowledge extracted from cadavers, Bell was able to compare his artistic creations with his experiences on the battlefield. After sketching 'extremity of pain' (Figure 10.3):

I had an opportunity to observe the truth of that expression. In extracting a bullet from the arm of a strong young man, I saw ...

Figure 10.3 Charles Bell, 'Anatomical Expression of Pain', *Essays on the Anatomy of Expression in Painting*, 1806 (Wellcome Library, London)

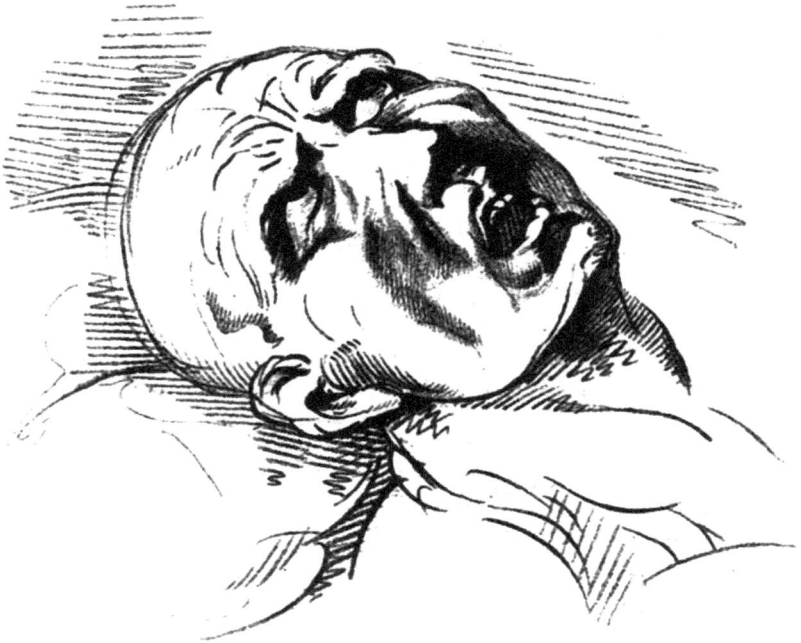

Figure 10.4   Charles Bell, 'Pain of the Sick', *Essays on the Anatomy of Expression in Painting*, 1806 (Wellcome Library, London)

that his face was turgid with blood; the veins on the forehead and temples distinct; the teeth strongly fixed, and the lips drawn so as to expose the teeth and gums; the brows strongly knit, and the nostril distended to the utmost, and at the same time drawn up.[9]

Preceding this illustration is one 'of bodily pain, anguish and death' (Figure 10.4), in which Bell notes a subtle difference:

In this plate the pain is that of the sick, and in some degree subdued by continual suffering. One striking feature of this expression is that of the confined nostril; for I have observed, that when the suffering does not approach to extreme agony, the nostrils are narrow and depressed.[10]

This further demonstrates how the expression of extreme pain is one that is difficult to maintain over a period of time. To understand the importance of the nostril for Bell, one must look to his comments in the revised version he published in 1824. The face is part of the wider nervous system, as Bell emphasises, and he believes the muscles around the nose are directly connected to the thorax. When confronted with the puzzling deaths of soldiers following amputations, Bell concluded that inflammation of the lungs was ultimately the cause of death and this highlights why he observed this area closely.

Bell concluded that the most expressive features of the face are the eyes and mouth as they are the most mobile. Anatomically there is a concentration of muscle structure towards these two areas. He also noted that distinctions between pain and other emotions can become blurred and intermingled. Bell cites Edmund Burke's comments in 'On the Sublime and Beautiful' of 1757 as an example:

A man in great pain has his teeth set; his eyebrows are violently contracted; his forehead is wrinkled ... Fear or terror which is an apprehension of death exhibits exactly the same effects.[11]

This draws from Bell the gentle riposte: 'Mr Burke in his speculations on fear, assimilates it, with perhaps *too little* discrimination to pain.'[12] However, Bell acknowledges that 'There cannot be great pain without being attended with the distraction of doubt and fears' and, in the case of extreme pain, a dread of death would naturally evoke a countenance of both fear and pain.[13] That singular emotional states may evolve into others or become a combination of several emotions has continued to frustrate those seeking to apply a reductive approach to facial expression.

In Bell's revised work of 1824, he prefaces the essays with references to Descartes and the influence of the mind upon the body, but he draws upon a quote from a Dr Beattie to establish that questions regarding how the soul, mind and body ultimately interconnect are answerable only to God (or 'the Creator' as written); to man, it will remain an unsearchable mystery.[14] But Bell believed it to be acceptable to use the talents bestowed by 'the Creator' to search and explain his 'works' and, even if he is not able to discover the connection between the emotions of the soul and those signs of the body, that the desire to comprehend the 'organs of expression' should not be extinguished. His dissections were a way of advancing knowledge, whereas to look only on the surface led to seeing the effects of passion as 'jumbled signs, quite incongruous, from an ignorance of their natural relations'.[15] For Bell, one needed to appreciate the internal systems and connections of the nerves in order fully to understand facial expression.

Photographing pain

Bell and Le Brun, the doctor and the artist, shared the tenet that 'passions' were 'natural' components of humanity that revealed themselves through the body and chiefly through the face, despite the two men's different purposes and contexts. But their examinations were largely based on visual observations, even in dissection. The emergence of electricity as a controllable stimulus on the living rather than the dead allowed Dr Guillaume Benjamin Amand Duchenne in the 1850s to experiment with muscle contraction under his direction and *control*. To add to this innovation, the availability of photography also allowed the results to be captured in ways never seen before. Duchenne's *Mécanisme de la physionomie humaine, ou, Analyse électro-physiologique de l'expression des passions*, published in 1862, was one of the first pieces of scientific research to include photographic illustration. In it he declares:

> If we were able to master the electrical current, an agent so analogous to the nervous fluid, and could limit its effect to individual muscles, we could certainly shed some light on their localised actions ... Armed with electrodes one would be able, like nature herself, to paint the expressive lines of the motions of the soul on the face of man. What a source of new observations![16]

For Duchenne, electricity was so close to the physiological processes of muscle control that its effects were a replication of 'nature' itself. Moreover, he stressed that it was a medium that he, comparable to an artist, could control at will. He was quick to note that by using this new method of 'localized electrization':

> The muscular contraction revealed their direction and their anatomical situation more clearly than the scalpel of the anatomist. At least this is true in the face, where one inevitably sacrifices, in anatomical dissections, the terminal portions of the muscle fibres that have their crucial insertion into the skin.[17]

Having experimented and refined his apparatus and techniques, Duchenne moved from cadavers to living subjects. In the early stages when the current was high, he required the services of a man with facial insensitivity who was immune to the painful procedure. He used moist electrodes so that the current could be easily regulated and did not burn the skin. With further development, he was able to achieve the same effect without a painful level of current passing through the muscle. The irony of the 'truthful' nature of Duchenne's expressions is that there is no emotion 'behind' them; the expression is brought about artificially. Whatever feelings we perceive the man to be experiencing (described as 'terror mixed with extreme pain') are absent. The interpretation of what emotion is expressed was necessarily one of invention. These were effects without affects: expressions devoid of the thing being expressed – the emotion itself. Duchenne was modelling the plasticity of the face and not 'creating emotions'. Again, it is ironic to note that he saw the use of this new technology as means to attain a degree of objectivity over the mechanical actions of the face.[18] Despite the technical advances of using electricity and photography, the interpretation of what emotion is expressed is a practice prone to subjectivity and argument. Duchenne created his own taxonomy of emotion and descriptors, virtually devoid of any explanation or justification as to how he came to translate particular muscular activity into certain emotions.

Duchenne's research focused on expressions that could be produced by single muscles and those that were the result of combined contractions. Pain, one of the six isolated muscle expressions that Duchenne listed, could be produced solely by the muscle which lowers the brows (Figure 10.5). But there are two further references to pain in the complex expression list: 'Terror, with pain or Torture' (Figure 10.6) and 'Great pain, with tears and affliction' (Figure 10.7). For Duchenne, the eyebrows alone denoted pain, but different degrees of pain, as well as pain related to other emotions (fear and sadness), involved other areas especially around the mouth.

One of Duchenne's expressed desires was to help artists' accuracy in portraying emotion. Returning to the Greek sculpture of Laocoon, he believed he had proved that the eponymous figure's forehead was an impossible piece of artistic fancy. He considered that the furrows created could only be produced by the action of *raised* eyebrows, associated with surprise and attention, and yet Loacoon's eyebrows were decidedly drawn inwards and downwards. Such simultaneous, contradictory actions, Duchenne argued, were physiologically impossible.

In capturing the results photographically, Duchenne admitted to many manipulations, edits and adjustments that helped him to deliver the effect he most desired, including props and positioning. An early decision to crop himself out of the original frame renders the architect of the scene largely invisible. Without a human presence shown directing the apparatus, we are only given a fragment of the experiment. The electric current was not perfectly constant and the sensitivity of the

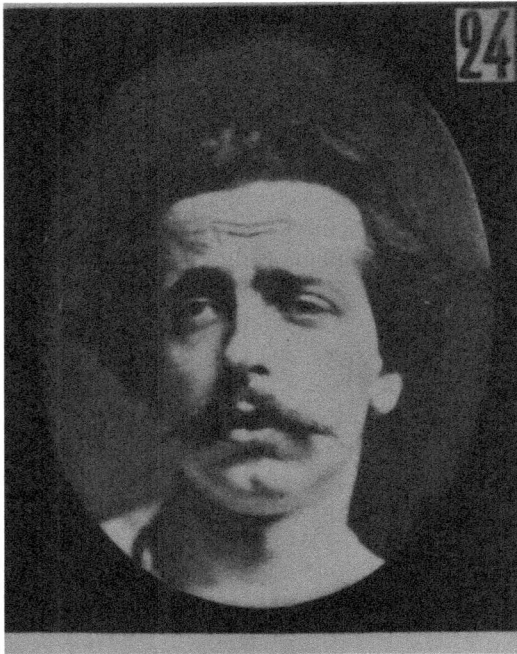

Figure 10.5   Guillaume Benjamin Amand Duchenne de Boulogne, 'Experiments in Physiology. Facial Expressions; Extreme Pain', 1862 (Wellcome Library, London)

muscle would weaken over time so that Duchenne would only uncover the lens when the muscle had reached the 'correct degree of contraction' necessary for the perfect expression 'that I wished to paint'.[19] He was choosing only to record the moments when he considered it appropriate. The image selected for publication is only one, then, of many potential images, yet the image is presented as singular and definitive.

For Duchenne, these artificially created contortions were simply elements of a language that could be understood by all cultures. The terms and language he used were never qualified or justified, except for an acknowledgement that a compilation from previous 'lists' of emotions by philosophers from Plato to Hobbes would produce a much greater range of vocabulary than he was prepared to consider. The interpretations of what the different muscular movements signified were based on his personal viewpoint. With mastery over these new tools, he could break down the 'written' emotions on the face into their component parts.

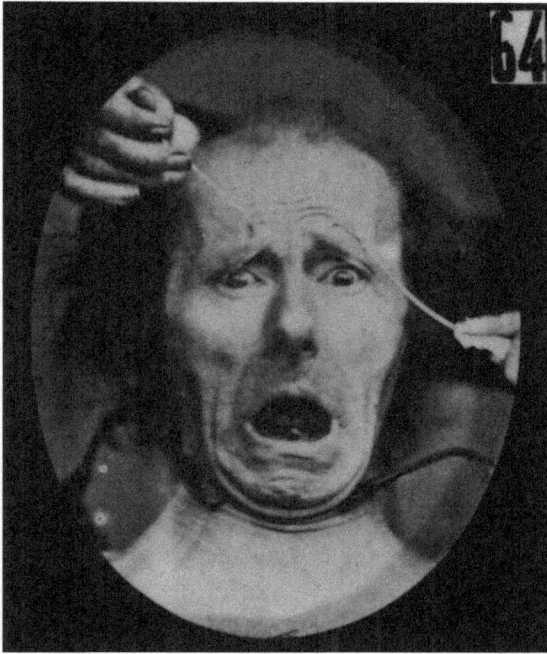

Figure 10.6 Guillaume Benjamin Amand Duchenne de Boulogne, 'Experiments in Physiology. Facial Expressions; Terror', 1862 (Wellcome Library, London)

Otniel Dror has asserted that the scientific prerogative in the early twentieth century was to turn the 'emotional interiority into a visually present, quantifiable, controllable, and rationalised object of knowledge' and what that partly relied upon was the elimination of alternative interpretations.[20] Duchenne in some ways was a pioneer of this approach. It was important to him to establish that the language of expression was fixed and innate. Since it was essentially 'silent', it would bypass the differences of the written and spoken word that separated nations, tribes and communities:

As man has the gift of revealing his passions by this transfiguration of the soul should he not equally be able to understand the very varied expressions successively appearing on the face of his fellow men? What use is a language one cannot understand? To express and to monitor the signs of facial expression seem to me to be inseparable abilities that man must possess at birth. The union of these two

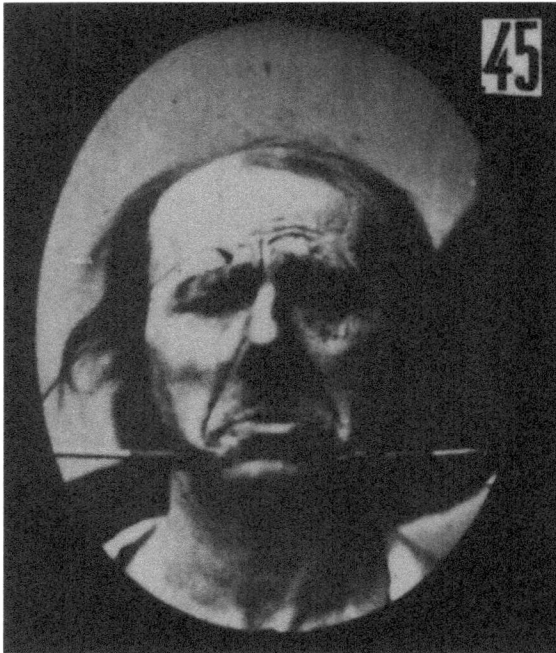

Figure 10.7   Guillaume Benjamin Amand Duchenne de Boulogne, 'Experiments in Physiology. Facial Expressions; Pain', 1862 (Wellcome Library, London)

faculties makes the play of facial expression a universal language. To be universal, the language must always be composed of the same signs ... [which] depends on muscular contractions that are always the same.[21]

Duchenne's argument for universality was, ultimately, contingent on a particular historical understanding of human nature. He blurs the distinction between his artificially produced expressions and 'actual' ones, noting that: 'The patterns of expression of the human face cannot be changed, whether one stimulates them or actually produces them.'[22] The revelation of 'passions' by a 'transfiguration of the soul' is, ultimately, indistinguishable from the mechanical manipulation of the face. The claim for universality depends, following Descartes, on the self-knowledge of the human soul. But how the emotional expressions of others are to be authenticated is actually rendered more uncertain than previously by Duchenne's own electrical innovations.

Scientifying pain

Duchenne's photographs were borrowed by Charles Darwin and, having been translated into illustrations, with all traces of the manual manipulation of the face removed, were used in his *The Expression of the Emotions in Man and Animals*, published in 1872. But Darwin departed from Duchenne's interpretation of the images he produced. A larger influence on Darwin's theorisation regarding expressions was his opposition to the 'design theology' that had underpinned the writings of Charles Bell. Thomas Dixon has concluded that Darwin did not think the primary or original function of expressions was to communicate a creature's inward mental state to its fellow creatures through outward signs. The communicative function was a fortunate additional outcome of the development of facial and bodily movements that originally had separate non-communicative purposes. Darwin, according to Dixon, explained these behaviours as inherited habits that were at some point in the past connected with the emotions of which they were now considered the 'expression'. Dixon notes that Darwin 'never attempted to explain the origins or function of emotional feelings nor did he try to define or classify the emotions per se'.[23] Instead, Darwin's main interest was in the physiology and behaviour associated with them.

Noting that frowning seemed to be a cross-cultural sign of perplexity when encountering something difficult or disagreeable, Darwin proposed that this action had its origin in infancy. A neonate's frequent response to any displeasing sensation or emotion was to scream, during which its eyes are strongly contracted. This, Darwin concluded, 'explains to a large extent the act of frowning during the remainder of our lives'.[24]

Essentially, these once-useful inherited habits had become, over successive generations, purposeless as environmental conditions changed. To undermine the theology of Bell's framework, Darwin had to deny that expressions were beneficial to the human race and evidence of divine design. As Dixon argues, if Darwin could show that these expressions were basically useless, Bell's arguments could no longer stand. For example, what practical use is blushing on account of shyness?

But whether these inherited habits could be shared by different cultures or agreed upon as signals associated with particular mental states led Darwin to conduct his own experiments on visitors to his residency at Down House, using both Duchenne's original photographs and several specially commissioned examples. Darwin asked his guests, mostly friends and acquaintances, to describe the emotion of an actor simulating grief in a photograph. Evidently Darwin was satisfied:

> That the expression is true, may be inferred from the fact that out of fifteen persons to whom the photograph was shown ... fourteen immediately answered 'despairing sorrow', 'suffering endurance', 'melancholy' and so forth.[25]

For Darwin, these words and phrases were unproblematic synonyms. What is significant is that this was an *actor* asked to portray a specific emotion. Like the subjects in Duchenne's photographs, the 'feelings' expressed were an artifice; they were not 'evidence' of an interior state. The viewers were being deceived by their perceptions and the image is offered without context in sublimation to a pursuit of 'objectivity'.

However, the responses to a photograph representing surprise were not so emphatic. Shown to a sample of 24 visitors, some added to the words 'surprise' or 'astonishment' the epithets 'horrified, woful [*sic*], painful, or disgusted'. These were 'widely different judgements', as Darwin admits in his introduction, acknowledging that some expressions were subject to more agreement than others. He attributes this discrepancy to 'how easily we may be misguided by our imagination', noting that by reading Duchenne's original accompanying text, he had a much better appreciation of the 'truthfulness' and accuracy of the results.[26] But the 'truthfulness' of these highly staged examples was based on agreement with Duchenne on what was being represented. Darwin added that 'if I had examined them without any explanation, no doubt I should have been as much perplexed ... as other persons have been'.[27] Darwin therefore re-evaluated his opinion after reading Duchenne's classification system and understanding the context and purpose for which the images were made. Without this context, attempts to interpret these images were subject to 'imagined' circumstances of their production and this resulted in confusion. Privileged knowledge, such as Darwin had, offered some further clarity.

Furthermore, Darwin was clearly influenced by Victorian conventions of gender roles and societal attitudes. His comment that 'the grief-muscles are brought into action much more frequently by children and women than by men. They are rarely acted on, at least with grown up persons, from bodily pain, but almost exclusively from mental distress' fitted into a broader scheme of Victorian gender codes that new scientific theories generally attempted to justify.[28] Despite the ubiquitous usage of the male image to depict universal expressions of pain and anguish, it was understood that reality would, more often than not, paint such expressions on the countenances of women.

Despite these methodological and cultural stumbling blocks, Darwin remained a believer that all the chief expressions of humanity were the same throughout the world. This he considered a further argument that the several races of our species descended from the same parent stock.

Authenticating pain

While Duchenne had used electricity artificially to stimulate expression and Darwin had included in his work photographs of actors, one Italian physiologist, Angelo Mosso, took the appeal to realism a stage further. Mosso has been described by Otniel Dror as a central figure in the physiology of emotions, whose extensive researches from the late 1870s created, measured, quantified and replicated emotions, and produced graphic and numeric representations of various affective states.[29]

Mosso was critical of Duchenne's approach: 'Classifications of mental faculties are too artificial', he argued, 'being derived from an abstraction based on facts and phenomena neither distinct nor definable.'[30] He went so far as to compare Duchenne's theory of localisations with Franz Gall's phrenological localisations. Gall exemplifies the subjectivity and awkwardness of language usage, especially when relating personality traits to physiological attributes, with respect to its socio-cultural context and its specific time and place. Gall's complex vocabulary may have once been 'fashionable', but it succeeded in obscuring rather than clarifying common human characteristics.

Mosso thought that pain demanded immediate and effective relief, and was eloquent in his description of how it could stimulate concern in others:

> Human pain is of such importance that all scientific curiosity becomes a trifling and ridiculous thing, and our mind rebels and feels an invincible repugnance to every desire which has not the alleviation of the sufferer for its object, to every act which does not spring from a lively and intense compassion.[31]

The use of photography was considered a major advance over artists' observations by Mosso. In an 1896 English translation of his book on

'Fear', he champions the suitability of 'instantaneous' photography and its profound effect on the viewer:

> No artist's fancy has ever been able to imagine or express what photography faithfully reproduces. In acute stages of suffering the human face inspires fear in one contemplating it; it is not alone the profound commiseration which we feel for the anguish of a sentient being which moves us ... but also the selfish thought that this palpitating flesh might be our flesh, that our soul, shaken with pain would also forget its tranquillity and our tortured nerves wring from us the same cries and the same tears.[32]

The strong emotional reaction Mosso describes in part hinges on the notion that this could easily be ourselves in this situation. The capacity to read the face of pain depends to some degree on the self-perception of fear. This was not empathy – an emotional identification – but a hypothetical reversal of positions. The confusion of this fear with compassion made the encounter with pain a complex situational drama. Mosso and his assistants were willing to put themselves through painful experiments in order to make their own observations, but this did not prove satisfactory for Mosso. The subjects voluntarily submitted to have their fingers squeezed against wooden inserts. Elements of anticipation, self-control and subdued responses were sufficient to alter the 'true' face of pain that Mosso was seeking to capture. He concluded that the facial expressions produced by the volunteers were less characteristic of the suffering he had witnessed elsewhere. In pursuit of greater verisimilitude, he left the confines of the laboratory and continued studying at Mauriziano hospital in Turin with real-life cases of injury. There he found an 18-year-old boy who had wounded his elbow. It had healed badly and was being treated, at intervals, by having the joint manipulated to promote mobility. The boy was photographed nearly every day for several weeks, whenever the surgeon forcibly extended the arm, which was intensely painful. The images were sequenced to show the differentiation between the fearful anticipation of the procedure and the actual point of physical pain (Figure 10.8).

Having criticised Duchenne for his rhetoric, Mosso rejects the possibility of discursive interpretation: 'I shall not attempt to describe these pictures, because I feel sure that no words of mine could express the transformation which the human face undergoes in pain.'[33] Without the paraphernalia, artificial stimulus and posed nature of Duchenne's images, Mosso's photographs offer a different level of

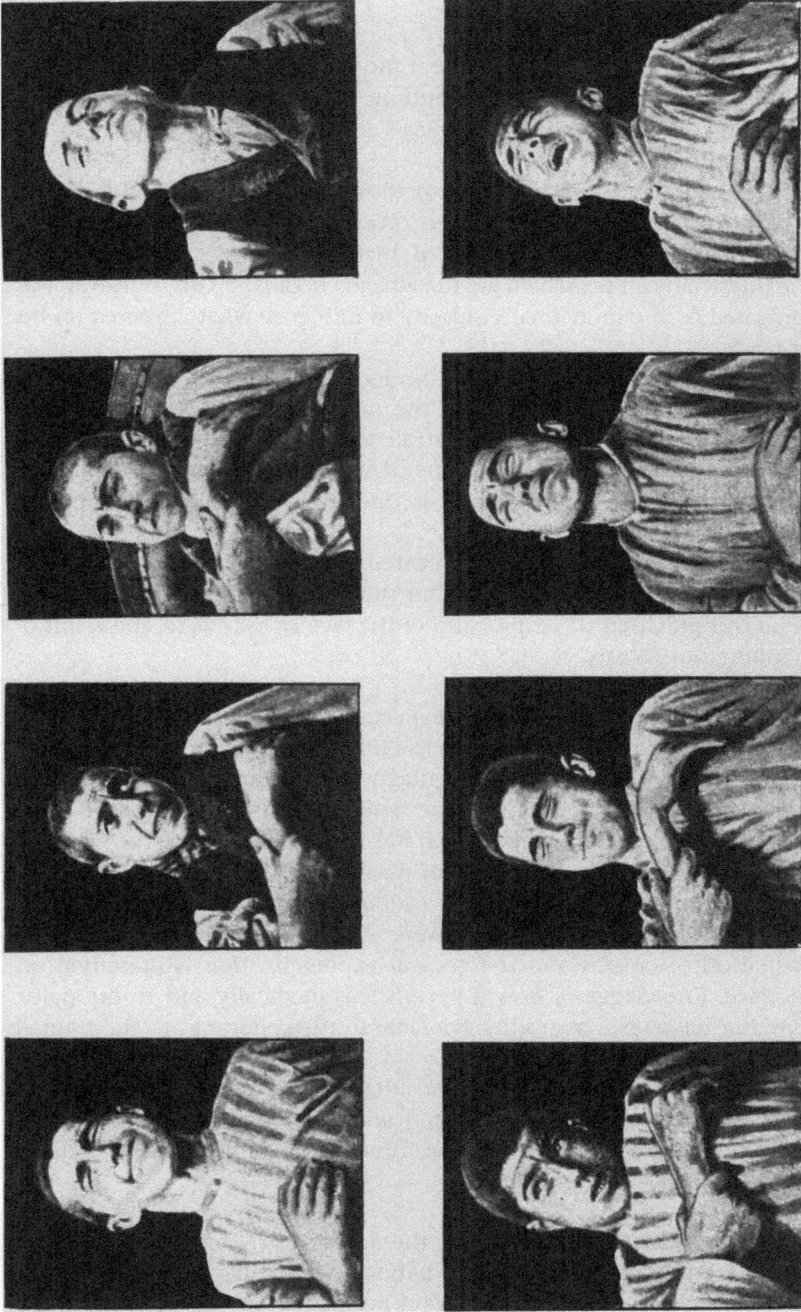

Figure 10.8 A. Mosso, 'Fear', *The Physiognomy of Pain*, 1868 (Wellcome Library, London)

visual communication. They appear more direct, more naturalistic and more personal. But this understanding of the vicarious experience of another's pain changed the emphasis from the subject of pain to the witness of pain.

Mosso had faith that medical experts' observations, guided by trained eyes, would get to the truth of pain: 'We may say that every malady has its peculiar expression of pain. Often, by merely looking at the patient ... the physician can tell which are the affected organs.'[34] This, however, was mitigated by the individual's capacity to influence what appeared on his or her face.

Mosso photographed patients undergoing operations without the use of chloroform and observed the range of ability to control their conduct. He attributed this variation to individual 'energy of will or weakness of character'.[35] Over time, it became apparent that pain was only one of a series of influences on our fluctuating emotional range:

> this study is very much complicated because of the rare occurrence of simple sensations of pain. Our states of mind are so variable and so complex that the expression of the face is, as it were, the result of numerous factors.[36]

He gives the labour pains of an expectant mother as an example: 'yet she finds a smile which expresses the hope of surviving'.[37]

Nevertheless, the belief in the diagnostic power of observation was shared by doctors themselves. One typical diagnostic manual of 1899, written by H.A. Hare, begins by reiterating how important the face and its expression is from the outset of diagnosis. The 'mental tendencies', habits, exposure to outdoor and indoor influences, and pathological processes going on in the body were all believed to contribute to the formation of creases and the general expression that is presented on the face. Knowledge of how a person felt, medically and emotionally, may not have been as readily available to the sufferer as to the trained observer (Figure 10.9).

In the intimate setting of the consultation, however, communication between doctor and patient was not unidirectional. The doctor was to be careful not to transmit any unintentional signals, as this helpful advice made clear:

> So much can be learned by the physician from the expression and general appearance of a patient's face that a careful inspection of these parts should always be made. For this reason, in the

Figure 10.9 Doctor Visiting a Sick Woman, c. 1800 (Wellcome Library, London)

consulting-room the physician should always arrange his chair in
such a way that the light falls upon the face of his patient, while
his own is in the shadow, and this is important not only because
the facial expression of the patient can thus be well seen, but also
because it prevents the patient from making a too close scrutiny of
the physician's face with the object of detecting encouragement, lack
of sympathy, or alarm.[38]

Hare believed it was not difficult to spot fake from genuine illness,
saying that the true facial expression of disease was rarely aped by a

malingerer. A genuinely ill patient lacked self-awareness regarding his or her appearance:

> It is not uncommon to see a person who is suffering from the onset of a sudden and grave disease bearing upon his face what we call an 'expression of anxiety' when he himself has no conception of the gravity of his illness. Furthermore in genuine cases, the individual will not recognise this when looking in the mirror.[39]

In strong contrast to the analysis of Darwin and Mosso, Hare attributed the importance of the emotional expression of pain to the advantage it gave to pain's witness, whose own emotional response could be trained as a diagnostic tool. In children not able to best describe the nature of their pain, some doctors believed that particular expressions could indicate where in the body it was occurring; a pinched nose showed that it was the chest and a raised upper lip meant the stomach. In adults, the upper lip drawn up in such a way as to show the teeth was associated with acute peritonitis, and twitching of mouth and eye muscles often accompanied pain below the diaphragm. In Hare's description of the condition known as 'paralysis agitans' (which probably corresponds to what we now would call Parkinson's disease), the face wore a 'distressed and pathetic, and yet somewhat intense' expression.[40] While acute pain could produce a transient contortion of the face, chronic pain, according to Hare, made a 'naturally gentle expression turn hard and stony, and in cases of pain in the head the expression is not only that of pain but of profound mental depression'.[41]

From their encounters with patients, doctors were prone to deliver their findings, with confidence, as the product of experience, with little recourse to objective measurement or data collection. This was in stark contrast to contemporaneous developments in the scientific recording of emotional data in the physiological laboratory, where the insight of the observer was replaced by the inscription of the machine.[42]

Conclusion

Although scientific approaches to the study of emotions strived to separate and isolate distinct states, the work of Duchenne and Darwin paid little attention to the distinction between reality and photography, and between 'genuine' emotional reactions and those simulated by actors or contrived by electrical apparatus. What is sacrificed in the reductionist approach to standardise and categorise emotional representation is the cursory attention paid to instances of misconception. As Boddice cautions: 'Where a smiling face leads us to conclude that happiness is occurring we are duped by the red herring of representation without context.'[43]

The subject of posing in these photographs and how the act of being observed might affect the 'self-consciousness' of the subject has been astutely commented on by Ruth Leys.[44] Leys also reflects upon the inability of photography to record the essential movement, over time, of facial changes. This is indicative of a relatively recent re-examination of the use of photographs as scientific documents and their validity.

In order to preserve the idea of objectivity of scientific observation, contextual information has been deliberately omitted from representations of emotions. This, however, is a crucial aspect of any encounter with emotional states and is especially pertinent to suffering. In the studies examined here, prioritisation has been given to the mechanisms of facial expressions, their actions and movements, their replication and structural organisation. The more complex and problematic considerations regarding interpretative processes of different expressions have not been explored to the same degree.

In Bell, Duchenne, Darwin and Mosso, respectively, the universality of the expression of pain has meant something subtly different according to each scientist's methods and purpose. With a plurality of universalities affecting the sufferer, the witness and the reader differently, and with the emphasis shifting among them, we should acknowledge that a range of historical meanings can be derived from a common (set of) expression(s).

Notes

1. R. Boddice, 'The Affective Turn: Historicising the Emotions', in Cristian Tileagă and Jovan Byford (eds), *Psychology and History: Interdisciplinary Explorations* (Cambridge University Press, 2014), 157.
2. C.E. Izard, *The Face of Emotion* (New York: Appleton Century Crofts, 1971), 188, emphasis in original.
3. K.M. Prkachin, 'Assessing Pain by Facial Expression: Facial Expression as Nexus', *Pain Research and Management*, 14(1) (2009): 54.
4. Signe Preuschoft, quoted in W.A. Ewing, *Face: The New Photographic Portrait* (London: Thames & Hudson, 2008), 190.
5. A.C. De C. Williams, 'Facial Expression of Pain: An Evolutionary Account', *Behavioral and Brain Sciences*, 25 (2002): 439.
6. C. Le Brun, *A Method to Learn to Design the Passions* (Berkeley: University of California Press, 1980 [1734]), 41.
7. *Ibid.*, 20.
8. René Descartes, *The Passions of the Soul* (1649).

9. C. Bell, *Essays on the Anatomy of Expression in Painting* (London: Longman, Hurst, Rees and Orme, 1806), 118.
10. *Ibid.*, 117.
11. *Ibid.*, 143.
12. *Ibid.*, emphasis added.
13. *Ibid.*, 144.
14. C. Bell, *Essays on the Anatomy and Philosophy of Expression*, 2nd edn (London: J. Murray, 1824), x.
15. *Ibid.*, xii.
16. G-B Duchenne de Bouogne, *The Mechanism of Human Facial Expression* (R. Andrew Cuthbertson (ed.)) (Cambridge University Press, 1990), 9.
17. *Ibid.*, 10.
18. For the seminal exploration of the construction of the objective image in science, see Lorraine Daston and Peter Galison, *Objectivity* (New York: Zone Books, 2007).
19. Duchenne, *Mechanism*, 106.
20. O.E. Dror, 'The Scientific Image of Emotion: Experience and Technologies of Inscription', *Configurations*, 7(3) (1999): 381.
21. Duchenne, *Mechanism*, 29.
22. *Ibid.*, 30.
23. T. Dixon, *From Passions to Emotions: The Creation of a Secular Psychological Category* (Cambridge University Press, 2003), 166.
24. C. Darwin, *The Expression of the Emotions in Man and Animals* (Joe Cain and Sharon Messenger (eds)) (London: Penguin, 2009), 205.
25. Darwin, *Expression*, 169.
26. *Ibid.*, 25.
27. *Ibid.*, 25.
28. *Ibid.*, 169. See R. Boddice, 'The Manly Mind? Revisiting the Victorian "Sex in Brain" Debate', *Gender and History*, 23(2) (2011): 321–40.
29. O.E. Dror, 'The Affect of Experiment: The Turn to Emotions in Anglo-American Physiology', *Isis*, 90(2) (1999): 205–37.
30. A. Mosso, *Fear* (London: Longmans, Green and Co., 1896), 187.
31. *Ibid.*, 201.
32. *Ibid.*, 202.
33. *Ibid.*, 203.
34. *Ibid.*, 204.
35. *Ibid.*, 203.
36. *Ibid.*, 204.
37. *Ibid.*
38. H.A. Hare, *Practical Diagnosis: The Use of Symptoms in the Diagnosis of Disease*, 3rd edn (London: Henry Kimpton, 1899), 25.
39. *Ibid.*, 27.
40. *Ibid.*, 30.
41. *Ibid.*, 27.
42. Dror, 'Scientific Image'.
43. Boddice, 'Affective Turn', 151.
44. R. Leys, 'How Did Fear Become a Scientific Object and What Kind of Object is it?', in Jan Plamper and Benjamin Lazier (eds), *Fear Across the Disciplines* (University of Pittsburgh Press, 2012).

11

'When I Think of What is Before Me, I Feel Afraid': Narratives of Fear, Pain and Childbirth in Late Victorian Canada

Whitney Wood

Private lives, emotional communities

In the late 1860s, while expecting her first child, Lucy Ronalds Harris, a young newlywed from a respectable London, Ontario family, confessed in her diary: 'I half fear that July [the month during which she was expected to give birth] will be the end for me ... I think I shall not recover.'[1] Although she did 'recover' after having her first child and ultimately went on to deliver four more children, her memories of fear and anxiety about giving birth marred her subsequent pregnancies. While in the first trimester of her fifth and final pregnancy in 1880, she still remarked 'when I think of what is before me, I feel afraid'.[2]

Harris' emotions during her pregnancy and feelings towards her coming delivery were representative of those of many women during the late-nineteenth and early-twentieth centuries. This chapter takes Harris' account, along with others found in the private writings of English-Canadian women, as a starting point to examine late-Victorian attitudes towards women's bodies and childbirth pain. As was the case in much of the Western world, the second half of the nineteenth century was a time of intensive socioeconomic change in Canada. As women became increasingly visible and active in the public realm, the middle class sought to preserve its status, immigration reached unprecedented levels, and gender, class and racial tensions intensified. Changing medical and cultural perceptions of the female body were one expression of these heightened social anxieties. During the mid- to late Victorian years, the trope of the 'delicate' middle-class white woman, highly 'evolved' and therefore increasingly sensitive to pain, became commonplace in both medical and public discourses. In the limited obstetrical training they received, the professional journals they read and produced, and the advice literature they published and promoted for young wives and expectant mothers, the majority of English-Canadian physicians articulated and supported perceptions of an increasingly delicate white female body. These doctors also asserted that this rapidly growing group of women, referred to by one physician as 'the luxurious daughters of artificial life', were apt to face new levels of pain in giving birth.[3] Giving birth, historically the heart of womanly culture, came to be shrouded in growing levels of fear and anxiety, paving the way for its increasing medicalisation and domination by male physicians.

English-Canadian women such as Lucy Ronalds Harris appear to have internalised these ideas. In the diaries and personal correspondence they left behind, they narrated their bodies and birthing experiences in ways that conformed to prevailing medical discourses. Medicalised descriptions of the 'delicate' female body and women's increased sensitivity both reflected and reinforced middle-class women's anticipatory fear of the pain of giving birth, and markedly shaped individual recollections of birthing experiences. These pervasive anxieties contributed fundamentally to some women's growing distrust of traditional female support networks and fuelled their increasing recourse to 'modern' physician assistance during the birthing process. The outcome was a different personal experience of pregnancy and birth for middle-class white women, the rapid elimination of midwife attendance and the burgeoning professionalisation of obstetrics in English Canada.

The private lives and experiences of women interested many early historians of emotion. One of those prominent in the field, William Reddy, has recently pointed out that the subdisciplines of the history of emotions and women's history evolved along parallel lines. In fact, some of the first researchers to become interested in the history of emotions were historians of gender and women, 'largely because women had always been considered more emotional than men'.[4] More recently, leading scholars in the field have conceptualised the history of emotions not as a distinct and separate specialisation, 'but as a means of integrating the category of emotion into social, cultural, and political history, emulating the rise of gender as an analytical category since its early beginnings as "women's history" in the 1970s'.[5] In other words, the history of emotions is an analytical tool – a particular 'way of doing' political, social and cultural history – rather than 'something to be added to existing fields'.[6] Focusing on the emotions is undoubtedly a valuable strategy for gender and women's historians, but scholars in these fields need to be mindful to avoid unintentionally highlighting and perpetuating the stereotype of women as hyper-emotional.

This study of English-Canadian women's private narratives of fear, pain and childbirth contributes to the still-embryonic historiography on emotion and pain by exploring one specific contextual example of the ambiguous relationship between the two. Though emotions such as fear, anger and happiness are universally central to the human experience – if not, however, universally experienced by individuals in different cultures – analysis of emotions has, as Joanna Bourke notes, 'remained peripheral to the historical discipline'.[7] As Bourke argues, the experiences of and rhetoric surrounding emotions such as fear are 'an expression of power relations. Emotions link the individual with the social in dynamic ways. They are always about social enaction'.[8] She points out that, of all the emotions, fear is fundamentally about and rooted in 'the body – its fleshiness and precariousness'.[9] Nineteenth-century fears of live burial, dissection and untimely and unrespectable death have transformed into twentieth-century anxieties surrounding the bodily pains and discomforts of disease, cancer and old age, but over the years, the body has remained a central site and focal point of human fears.[10]

Late nineteenth- and early twentieth-century anxieties about pain and parturition fit into this framework of fear. In their shared fears and anxieties about the pain of giving birth, middle-class white women formed what Barbara Rosenwein classifies as an 'emotional community' – a group of people 'animated by common or similar interests, values, and emotional styles and valuations'.[11] Like other proponents of the cognitive theory of emotions, Rosenwein argues that emotional responses 'are about things judged important to us' and 'are the result of our values and assessments'.[12] Women's fears during pregnancy and anxieties surrounding an upcoming birth, then, are a reflection of the significance of a maternal identity for women during the Victorian period.[13] This significance also ascribed a unique meaning to labour pain. Effecting a visible change on the female body – and therefore, according to Elaine Scarry, all the more likely to be treated – the pain of giving birth increasingly represented, in the words of one physician, 'a public health question' and an obstacle that prevented white, middle-class women from fulfilling their maternal true purpose.[14] In so doing, this pain threatened the health and vitality of the Anglo-Canadian race.

Women's diaries and personal correspondence offer a particularly valuable viewpoint into private and individual experiences of fear, anxiety and pain within this particular emotional community, but these types of sources have their own well-worn methodological considerations. Scholars have argued that women's personal narratives suggest 'how women negotiate their "exceptional" gender status both in their daily lives and over the course of a lifetime'.[15] Historians, in particular, consider that diaries provide rare 'accounts of domains that need to be better understood' and are the most important window into women's experiences of some of the 'key moments' of domesticity, including birth.[16] Yet these documents also 'recount a process of construction of the self' and, accordingly, are always mediated to a certain extent by the diary-keeper.[17] This personal mediation shaped the entries in the individual diaries discussed in this chapter for some women more than others. Lucy Ronalds Harris' stepmother, Amelia Ryerse Harris, for example, was well aware that her diary had an audience, was read by relatives and often served as a forum for wider family discussions.[18] For other women, diary entries were shaped by time, as they 'wrote backwards' to make up for weeks or months of missed diary-keeping after events such as delivering a child, or treated their journals as the site for year-end recaps of goings-on in the family.[19] Women's memories of pregnancy and parturition were also, of course, mediated by the pain of giving birth. As one scholar has recently argued, traumatic events – including, I would suggest, a particularly distressing or painful experience of childbearing – are characterised by their 'inability to be integrated into one's normal patterns of meaning-making' as memories are moulded by traumatic experience.[20]

Aside from intercessions by the diary-keeper, diary writings were also mediated by the wider cultural milieu, and scholars have aptly identified that diary writings are 'materially and socially situated'.[21] Diary entries, then, are not simply reflections of lived realities, but, rather, offer 'nuanced commentaries on the cultural context[s] in which women were required to function'.[22] On a more basic level, these sources typically speak to the views of a particular group of women. Keeping a diary demanded basic literacy, which, throughout the nineteenth century, excluded most members of the lower classes. In addition, diary-writing during the late Victorian years was 'associated with a genteel life and an ideology of refinement'.[23] The practice became a way to 'indicate class standing' and tended to exclude most members of the working class as well as middle-class men who worked outside of the home. Diary-writing during this period, for the most part, 'marked women of leisure'.[24]

As a result of the limited nature of diary-keeping during the late Victorian era, the diaries examined in this chapter are all written by white, English-Canadian women who could be easily identified as members of a 'respectable' middle class.[25] These women, however, as was typical during this period, were generally reticent to discuss pregnancy and childbirth.[26] This reluctance may have stemmed from a variety of factors – women would have been aware of the potential and realities of losing a pregnancy in the early stages; discussing reproduction and pregnancy, even in a diary, may have pushed the limits of Victorian feminine respectability; and, in the eyes of many nineteenth-century physicians, pregnancy was considered to be a 'normal' and unremarkable state for the married woman.[27] The result is that oftentimes, only cursory references to these major life events made their way into women's private writings.

Despite these limitations, Judith Walzer Leavitt, a prominent American historian of childbirth, has argued that fear of childbirth was a common anxiety for women during the nineteenth and early twentieth centuries: 'Rich, poor, urban and rural women all shared with each other, by virtue of their sex, an enormous bond of common experience ... Owing to their common physical and social experience, women developed similar feelings, fears, and needs during pregnancy and delivery, despite their divergent life circumstances.'[28] While Walzer Leavitt's study of American women's perceptions of childbirth is now nearly 30 years old, no Canadian-focused equivalent exists.

Fears of pain, pains of fear

In late nineteenth- and early twentieth-century English Canada, women's fear of the pain and suffering associated with giving birth was often accompanied by their fear of invalidism following delivery and, more importantly, the fear that they or their infants would face death in the birthing room. These anxieties made the whole pregnancy a particularly tense time for many women. The fact that these emotions tended to characterise the entire nine-month period goes hand-in-hand with the expansive definitions these women had for the pains and discomforts of pregnancy and the act of giving birth. For the women authors of these diaries, the pain of giving birth *was* emotional and often inseparable from the wider uncertainties, anxieties and fears of the entire duration of the pregnancy. Lucy Maud Montgomery, the internationally known children's author made famous by her *Anne of Green Gables* stories, best encapsulated this relationship. For Montgomery, the many uncertainties associated with childbearing, coupled with the increasing bodily discomforts of gestation, led her to declare in the final weeks of her first pregnancy in 1912: 'I really suffer a martyrdom of misery, partly physical, partly anxious.'[29] This 'misery' extended well beyond the actual birth and arose out of both her anxieties and corporal complaints, blurring the lines between emotional and physical sufferings. In consistently discussing the pain and discomforts of pregnancy and childbirth using emotionally charged language, Montgomery and the other diarists challenge what David B. Morris has referred to as the 'Myth of Two Pains' – the idea that pain can be logically divided into two separate types: physical, and emotional or mental.[30] The remainder of this chapter traces English-Canadian women's emotional responses to the discomforts of gestation and childbearing over the course of the pregnancy and, in so doing, contributes to a growing body of scholarship that seeks to collapse the artificial divisions between emotional and physical pain.

Though many of the women discussed in this chapter tended to frame their pregnancies as being times shaped by fear, uncertainty and anxiety, it is important to point out that these 'negative' emotions were often tempered by 'positive' feelings of excitement and joy towards the birth of a new child. While these feelings prompted some women, including Montgomery, excitedly to record the discovery of a pregnancy in a diary or journal, veiled references to pregnancy and later allusions to the fact that a child might be expected in the near future were more common.[31] Frances Tweedie Milne of Whitby, Ontario, for example, made note in the margins of her diary in January 1870 that 'courses should have come on last' before Christmas the previous month. In mid-February, she made a similar note that she 'went to church … courses should have cm [come]'. During her second trimester, references to the 'little gown' and 'little clothes' she sewed and showed to a friend suggested the coming addition to the family.[32] Similarly, western Canadian homesteader Eliza Jane Wilson commented that she 'found the basin in the bedroom empty and clean … rather unusual' in late August 1903, but did not admit to 'getting … ready for the new addition' until 30 November.[33]

The fear of losing a pregnancy in its early stages might have contributed to such self-censorship, but it also prompted other women to record the anxieties that surrounded a new pregnancy. After miscarrying in the spring of 1919, Gwyneth Logan of Vancouver, British Columbia wrote to her husband Harry who was working in Ottawa, Ontario in the autumn, during her first trimester: 'I can't help feeling nervy after last time. I shall be thankful when the next month is safely over … these are critical days … I can't help being anxious.'[34] The precariousness of pregnancy meant that Logan, like other women during the period, often trod carefully during these months. After experiencing repeated spotting and going on bed rest, Logan complained and described herself as 'a semi-invalid'.[35] For women who may have had more considerable domestic responsibilities, the often doctor-prescribed rest associated with pregnancy could have posed a problem.

English-Canadian women also expressed distaste with some of the other physical discomforts associated with pregnancy. Lucy Ronalds Harris wrote of her lack of energy and feeling 'unfit for everything except bed' while pregnant with her first child in January 1868.[36] During a subsequent pregnancy, she made note of her 'many vexations' and remarked that 'no one knows what I suffer'.[37] Eliza Wilson repeatedly mentioned suffering from toothaches and neuralgia before, during and after her pregnancy,[38] demonstrating that oftentimes, the various 'ailments' associated with pregnancy were, for many women, part of a wider spectrum of longer-term health complaints.

Historians have demonstrated that as the traditional culture of social childbirth declined in the nineteenth century, giving birth became increasingly 'shrouded in mystery' for many American young women.[39] An examination of Canadian women's private records demonstrates that the uncertainty associated with childbirth was undoubtedly a major cause of anxiety for women north of the border. Pregnant with her first child in 1912, Lucy Maud Montgomery remarked that she felt very nervous when she thought of the ordeal before her, speculating that 'it cannot be easy at the best'.[40] For her, the uncertainty associated with birth was one of her greatest fears. After her delivery, she recounted:

in the dead, dim hours of night, fears and gloomy dreads came to me ... they always lurked in the background of my mind. Would I escape with my life? Would I, as some of my friends have done, suffer so dreadfully that the remembrance would always be a horror? Would my child live? These and a score of other fears haunted me.[41]

Somewhat confident that the threat of miscarriage had passed after the end of her first trimester, Gwyneth Logan wrote to her husband: 'I don't believe I shall have any more troubles, but one never knows, that is the worst of this business.'[42] As women approached birth anxious, uncertain and oftentimes largely ignorant about their own bodies, physicians' perceptions 'became the dominant view of the nature of birth', and as many physicians were undoubtedly less than comfortable in their new role, their actions and perhaps unsympathetic demeanour at the bedside may have well exacerbated women's anxieties.[43]

Given contemporary medical ideas about the delicate female body, many of the English-Canadian diarists also noted their fear of the unknown and reputedly unprecedented pain they would experience in giving birth. The prescriptive literature of the period emphasised that primipara mothers had the longest and most difficult deliveries, and so first-time mothers had reason to suspect that they had the most to fear.[44] Having had no previous children, Montgomery – who, at 37, was well over the age of 26 after which a woman was considered an 'older mother' subject to increasingly complicated (and painful) deliveries[45] – reasoned that the discomforts of pregnancy would be a small price to pay 'if it were not for the anguish of the final ordeal'. She continued: 'I have never had to endure any intense physical pain. So I fear I shall not bear it well or be very brave or patient ... now, when the end is coming so near, I cannot avoid feeling dread and anxiety.'[46] Montgomery's statement about her suspected lack of stoicism in the face of new levels of pain is particularly interesting. As Lucy Bending has suggested in her study of the representation of pain in nineteenth-century England, medical men commonly asserted that 'supercivilised' women (of Montgomery's sort) were particularly apt to 'eschew all painful encounters', behaviour which, in the long run, was thought to weaken the female body and render it more susceptible to pain.[47] As this line of reasoning suffused the medical discourse of the period, Montgomery may well have been aware of such arguments and they would have undoubtedly heightened her anxiety.

In her study of childbirth and anaesthesia in the USA, Jacqueline Wolf argues that by the mid-nineteenth century, women's fear of pain during childbirth had equalled their fear of death during the ordeal, but that the two fears consistently converged as 'women often likened labour pain to a near death experience'.[48] This perspective also comes through in the writings of English-Canadian women. In the months before her first delivery in the summer of 1868, Lucy Ronalds Harris, whose account opened this chapter, remarked that she feared giving birth would result in her death.[49] Such fears were nothing new for the Harris women. After losing one daughter to puerperal fever, Lucy's mother-in-law, Amelia Ryerse Harris, remarked in her diary eight years earlier that she feared for the life of another, who expected her confinement within days.[50] Montgomery, who wrote at length on her fears of the pain of giving birth, also contemplated her fate, wondering: 'Will I pass safely through the valley of the shadow and bring therefrom a new life? Or shall I remain among the shadows? I shall not write in this journal again until all is over. Perhaps I may never write again...'[51]

Seeking succour in the medical profession

To counter some of the vulnerability associated with pregnancy and assuage some of these fears, women traditionally sought out their own mothers and other female family members and friends who had survived the experience of giving birth. By the second half of the nineteenth century, however, many of these networks were being disrupted by increasing migration and family mobility. Writing from the west coast in 1853, Georgina Bruce Kirkby described her pregnancy in the apparent absence of female kin and company, writing that if anything 'could relieve or comfort me under my present very depressing condition of health ... it would be a congenial female companion. Every woman needs a companion of her own sex'. She later recalled a two-day visit from a neighbouring woman, 'which has quite made me forget myself and my ailments'.[52] Canadian women also made note of these disruptions. Mary Kough Brown, of Hamilton, Ontario, who had immigrated from England shortly before her 1865 pregnancy, recorded 'a most depressed day, sick of waiting and longing for such comfort as only a dear mother could give'.[53] Frances Milne, who noted her 'missed courses' in the margins of her diary, wrote in the month before her delivery that she was 'looking for Whitby people' arriving at the train station, 'but disappointed'.[54] Gwyneth Logan wrote frequently of her letters home to her mother in England and consistently reminded her husband in Ottawa that their separation during her anxiety-ridden pregnancy was 'an additional trial'.[55]

Judith Walzer Leavitt has argued that women's fear of childbirth, pain and death 'eroded the comfortable feelings that women received from their companions during traditional births' and, in part, contributed to their greater willingness to seek out physician assistance.[56] In late nineteenth- and early twentieth-century English Canada, when these traditional womanly support networks were increasingly fractured or conspicuously absent, middle-class women were all the more open to the new possibilities offered by physician-assisted birth. For some women, this even meant going to hospital for the birth, well ahead of the general Canadian trend.[57] With a husband frequently away in Calgary on 'ranching duties' and after a previous miscarriage, Eliza Wilson, for example, remarked: 'I have decided to go to the Hospital on Friday. I hope everything will be all right this time.'[58] Being in a hospital setting may have done little to calm Wilson's nerves. Even after being admitted, she wrote of her continuing hope that everything was 'all right', and in a statement that again blurs the boundaries between emotional and physical pain noted that the experience 'scares a lady nearly to death'.[59]

As a rural homesteader and recent newcomer to the area, Wilson may have had few alternatives to the hospital. Canadian historian Wendy

Mitchinson has pointed out the agency of most women in their encounters with physicians and hospital-based births. On the most basic level, 'women often controlled whether and when they would see a physician', and it is important to keep in mind that 'they sometimes supported the increased medicalization of their lives'.[60] When women, influenced by the prevailing discourses of the period, sought out physicians and hospital-based births that were represented and perceived as safer and more 'modern', 'women's agency allied itself with the medical profession'.[61]

However much anxieties and expectations of childbirth pain may have been influenced by medical rhetoric that held that 'for the great majority of women in civilized nations, parturition is a period of intense pain', women did and continued to have varying reactions to the unique, private and highly subjective pain of giving birth.[62] For many, physicians' predictions of the heightened pain they were expected to experience were brought to bear, and they recalled their confinements as times coloured by both pain and fear. Mary Kough Brown noted that 'a fat healthy boy' was born to her only 'after much suffering'.[63] Nearly 40 years after the 1894 birth of her first daughter, Alberta homesteader Evelyn Cartier Springett recalled that 'during one long horrible night I suffered in silence ... I shall never forget those awful hours'.[64] Other women were surprised and thankful that they did not experience the pain they anticipated. Though Montgomery wrote at length of her fears, after giving birth, she admitted suffering 'many more a night with toothache ... I have had my baby and none of my forebodings have been fulfilled. I can smile at them now – but they were nonetheless harrowing while they lasted'.[65]

Despite this relief, Montgomery and many other mothers continued to fear their subsequent deliveries. And for those who had had unpleasant birthing experiences, childbirth understandably continued to be a cause for anxiety. One American woman remarked in 1871 that she dreaded a second birth 'with a dread that every mother must feel in repeating the experience of childbearing'.[66] Similarly, Jennie Curran of Orillia, Ontario noted in 1877 the worries that surrounded her pregnancy and growing family.[67] Lucy Ronalds Harris, after 'recovering' from her first birth, wrote again of her fears of dying while having her second child, and in her fifth and final pregnancy over a decade later, she still 'felt afraid' when she thought of what awaited her in the birthing room.[68] And despite her explicitly stated relief after her first birth, Montgomery noted while pregnant with a second son two years later, 'somehow I look forward to this second birth with more anxiety ... perhaps because I realize more clearly how many things might go wrong'.[69]

Women's persisting, and, perhaps, increasing anxieties went hand-in-hand with the growing pathologisation of both pregnancy and childbirth. During the second half of the nineteenth century, pregnancy was increasingly cast as a medical 'condition' and as a time when so-called 'female troubles' could reach new and unprecedented heights.[70] Physicians, who articulated the symptoms leading to the diagnosis of pregnancy, described various treatments for its many associated ailments and prescribed 'parturient balms' to all women 'for the purposes of rendering childbirth more easy', addressed these issues at length in the medical advice literature of the period, and women both contributed to and echoed this language when describing their pregnancies and confinements.[71] In noting that she expected soon 'to be laid aside' in the winter of 1879, Jennie Curran wrote that she hoped her 'illness' would 'be made a blessing'.[72] Throughout her childbearing years, Lucy Ronalds Harris repeatedly referred to her several pregnancies as her 'troubles' and informed her mother-in-law to expect that she 'would be ill' come the month of an anticipated confinement.[73] In 1919, Gwyneth Logan readily admitted her belief that pregnancy 'was bound to be a troublesome time ... even under the most favourable conditions'. Referring to her state as her 'condition', she quipped: 'I certainly am having my full share of symptoms this time.'[74] Echoing and fuelling medical rhetoric that consistently equated 'sickness' with the beginning of a labour and confinement, mothers also pathologised the actual birth, recalling and referring to it as an acute period of disease or exceptionally ill health.[75] Frances Milne noted that she was 'very sick all night' when her first child was born.[76] Constance Kerr Sissons of Fort Frances, Ontario described feeling 'very ill' before and during the birth of her first daughter in 1903.[77] Likewise, Eliza Wilson recalled that she was 'taken sick' before 'a nice strong lassie' was born to her in 1904.[78]

Physicians and mothers alike expected pain to go hand-in-hand with the 'termination' of the pathologised condition that was pregnancy, and the diarists discussed in this chapter stated that ideas about the relationship between pain and childbirth were 'common sense'. Again, Montgomery perhaps put it best:

> I have heard much about the agony of the birth chamber. That such agony is the rule rather than the exception generations of suffering women have testified since the dawn of time ... All my life I had heard and read of the anguish of childbirth, its risk, its dangers. There were times when I could not believe I would get safely through.[79]

Conclusion

Though the experience of pain is private, individualised and highly sub-jective, these shared anxieties had a real impact on parturient women in turn-of-the-century English Canada. As Jacqueline Wolf has pointed out, contemporary studies on birthing pain suggest that women's prior anxieties and expectations of severe pain are factors associated with the most painful labours and deliveries.[80] It is clearly impossible to determine the actual levels of physical pain that these 'delicate' women experienced in giving birth. Recurring emphases on the common anxie-ties and fears that surrounded parturition nonetheless demonstrate that for this group of women, the suffering associated with giving birth had a carefully articulated and distinguishable emotional component. These shared anxieties influenced the medical choices of this particular 'emo-tional community,' increasingly prompting middle-class women to seek out physicians for the scientific and 'modern' comfort they could offer in their interventions, including anaesthesia. Women always managed to retain some agency in this process. Though Wendy Mitchinson has demonstrated that 'only rarely in the debate over intervention were the demands of women heard', it is important to keep in mind that obstet-rical interventions were not always imposed on women against their will.[81] Women also retained agency in more subtle ways. Nancy Theriot has argued that 'women patients were active participants in the process of medicalizing *woman*' as a gender category and identity.[82] In this case, by both conforming to and contributing to existing medical dis-courses, women were active participants in the process of medicalising childbirth as well.

Notes

1. Diary of Lucy Ronalds Harris, 28 February 1868, in Robin S. Harris and Terry G. Harris (eds), *The Eldon House Diaries: Five Women's Views of the Nineteenth Century* (University of Toronto Press and the Champlain Society, 1994), 424.
2. Diary of Lucy Ronalds Harris, 24 March 1880.
3. M.L. Holbrook, *Parturition without Pain: A Code of Directions for Escaping from the Primal Curse* (Toronto: Maclear and Company, 1875), 17.
4. William Reddy, in Jan Plamper, 'The History of Emotions: An Interview with William Reddy, Barbara Rosenwein, and Peter Stearns', *History and Theory*, 49 (May 2010): 238.
5. *Ibid.*, 237–65.
6. Reddy, in *ibid.*, 249.
7. Joanna Bourke, 'Fear and Anxiety: Writing about Emotion in Modern History', *History Workshop Journal*, 55 (2003): 112–13.

8. *Ibid.*, 113.
9. Joanna Bourke, *Fear: A Cultural History* (London: Virago Press, 2005), 8.
10. Bourke, 'Fear and Anxiety', 118; and *ibid.*
11. When she initially coined the term in 2002, Rosenwein defined 'emotional communities' as 'precisely the same as social communities – families, neighborhoods, parliaments, guilds, monasteries, parish church memberships', and went on to suggest that researchers examining such communities seek to uncover 'systems of feeling: what these communities (and the individuals within them) define and assess as valuable or harmful to them; the evaluations that they make about others' emotions; the nature of the affective bonds between people that they recognize; and the modes of emotional expression that they expect, encourage, tolerate, and deplore', Rosenwein also asserted the fluid nature of these communities and argued that 'people moved (and move) continually from one such community to another ... adjusting their emotional displays and their judgements of weal and woe to these different environments'. Barbara H. Rosenwein, 'Worrying about Emotions in History', *American Historical Review*, 107(3) (2002): 842. Rosenwein offered a more succinct definition, stating that emotional communities are 'social groups that adhere to the same valuations of emotions and how they should be expressed' or 'groups of people animated by common or similar interests, values, and emotional styles and valuations' in a 2010 interview with Jan Plamper. Barbara Rosenwein in Plamper, 'The History of Emotions', 253.
12. Rosenwein in *ibid.*, 251.
13. The fact that motherhood existed at 'the apex of a Victorian bourgeois ideal' is a point that has been identified by a host of historians of women's health, sexuality and childbirth. See, for example, Hilary Marland, 'Languages and Landscapes of Emotion: Motherhood and Puerperal Insanity in the Nineteenth Century', in Fay Bound Alberti (ed.), *Medicine, Emotion, and Disease, 1700–1950* (Basingstoke: Palgrave Macmillan, 2006), 53. Though Marland's study focused on England, motherhood was central to middle-class femininity for women in Victorian Britain, Canada and the USA.
14. Scarry asserts that 'given any two phenomena, the one that is more visible will receive more attention'. Accordingly, it can be argued that the visibility of labour pain prompts the medical profession to pay more attention to its treatment, particularly as anaesthesia was most often given during the final stages of delivery, when the most visible changes took place. See Elaine Scarry, *The Body in Pain: The Making and Unmaking of the World* (New York: Oxford University Press, 1985), 12. Some physicians publicly emphasised connections between pain relief and eugenics. In a 1916 article, Dr J.C. Edgar suggested that 'eventually an established method of painless labour may be considered among public health questions [and] in the future may limit birth control and criminal abortion': 'Painless Labor', *The Canada Lancet*, 50 (1916): 218.
15. The Personal Narratives Group (ed.), *Interpreting Women's Lives: Feminist Theory and Personal Narratives* (Bloomington: Indiana University Press, 1989), 4–5.
16. Harris and Harris, *The Eldon House Diaries*, xxi.
17. The Personal Narratives Group, *Interpreting Women's Lives*, 5.
18. Harris and Harris assert that this diary in particular, rather than offering a mere personal record of events and private thoughts, 'played a didactic role

and was intended to instruct, sensitize, and even direct family members according to Amelia's wishes'. In her diary entries, Amelia Ryerse Harris made several comments to this end, writing most explicitly on 4 June 1859: 'My journal came under discussion.' The fact that Harris knew her diary was read by others and was, in effect, 'up for discussion' undoubtedly shaped the nature and tone of her entries. Harris and Harris, *The Eldon House Diaries*, xxv, 113. Kathryn Carter has argued that many diaries were, in fact, 'semi-public documents. They did (and do) circulate. The relationship between diaries and their deliberate or accidental audiences sets diaries apart from other forms of published writing'. Kathryn Carter, *The Small Details of Life: Twenty Diaries by Women in Canada, 1830–1996* (University of Toronto Press, 2002), 13.

19. After a period with no diary entries, Eliza Jane Wilson of Western Canada wrote on 9 January 1904: 'I have quite a diary write up. Almost 3 weeks I will write backwards, it will be easier.' Diary of Eliza Jane Wilson, 9 January 1904, Eliza Jane Wilson Fonds, 1901–1958, M 1320, Glenbow Museum and Archives. At the end of 1878, Jennie Curran of Orillia, Ontario wrote that 'another year is near its close and I think of my old friend and long to express my thanks to god for the mercies of another year'. Curran's particular habit of year-end entries speaks to the blurred and ambiguous nature of women's private writing practices, as she tends to treat her diary more as a memoir and a place to record exceptional life events after they have occurred. Diary of Jennie Curran, 28 December 1878, JEG Curran Fonds, MG 30 C85, Library and Archives Canada.

20. Michael S. Roth, *Memory, Trauma and History: Essays on Living with the Past* (New York: Columbia University Press, 2012), xviii.

21. Carter, *The Small Details of Life*, 22.

22. Harris and Harris, *The Eldon House Diaries*, xxii.

23. Carter, *The Small Details of Life*, 15.

24. *Ibid.*

25. I came to this conclusion based on individual entries referring to various elements of what can only be described as a 'comfortable' lifestyle – descriptions of frequent parties, extensive travel, elaborate houses and reference to domestic workers were the norm for these women.

26. In her examination of the American context, Jacqueline Wolf suggested that many women did not mention a pregnancy until after giving birth. I have found the same to be true for some Canadian women, but their writings are, understandably, of less interest in this chapter. Jacqueline Wolf, *Deliver Me from Pain: Anesthesia and Birth in America* (Baltimore: Johns Hopkins University Press, 2009), 31.

27. In her study of the nineteenth-century medical practice of Ontario practitioner Dr James Miles Langstaff, Jacalyn Duffin suggests that the fact that 'pregnancy seems to have been considered the normal state for a married woman' contributed to the doctor's lack of discussion of obstetric cases in his patient casebooks. Jacalyn Duffin, *Langstaff: A Nineteenth-Century Medical Life* (University of Toronto Press, 1993), 183.

28. Judith Walzer Leavitt and Whitney Walton, '"Down to Death's Door": Women's Perceptions of Childbirth in America', in Judith Walzer Leavitt (ed.), *Women and Health in America: Historical Readings* (Madison: University of Wisconsin Press, 1984), 155–65 at 156.

29. Mary Rubio and Elizabeth Waterson (eds), *The Selected Journals of L.M. Montgomery, Volume II: 1910–1920* (Oxford University Press, 1987), 30 April 1912.

30. David B. Morris, *The Culture of Pain* (Berkeley: University of California Press, 1991), 9. See also David Biro, Chapter 4, this volume.

31. Montgomery excitedly and famously proclaimed on 28 January 1912: 'I want to have a child ... something to link me with the future of my race.' *Selected Journals of L.M. Montgomery*.

32. Diary of Frances Tweedie Milne, entries 14 January 1870, 13 February 1870, 9 June 1870, 11 June 1870 and 26 August 1870, Frances Tweedie Milne Papers, MU 866, Archives of Ontario.

33. Wilson was living near Calgary in Dorothy, Alberta. Diary of Eliza Jane Wilson, entries 27 August 1903 and 30 November 1903, Glenbow Museum and Archives.

34. The Logan correspondence is a remarkable source, detailing the courtship and married life of Gwyneth (Murray) Logan and her husband, Harry Logan, for a period of more than a decade. Letters from Gwyneth Logan to Harry Logan, 6 October 1919 and 28 October 1919, Harry Tremaine Logan and Family Fonds, MG 30 C215, Library and Archives Canada.

35. Gwyneth Logan to Harry Logan, 29 November 1919, LAC.

36. Diary of Lucy Ronalds Harris, 22 January 1868, *The Eldon House Diaries*.

37. Diary of Lucy Ronalds Harris, Entries, 22 May 1873 and 24 May 1873. *The Eldon House Diaries*. These entries were written in the final months of Harris' pregnancy; her first son was born on 15 June 1873.

38. Diary of Eliza Jane Wilson, Entries 7 October 1901, 19 November 1901, 28 November 1901, 30 November 1901, 23–25 February 1903, 2 October 1903, 27 November 1903, 18 April 1904 and 25 May 1904, Glenbow.

39. Wolf, *Deliver Me from Pain*, 20.

40. *Selected Journals of L.M. Montgomery*, 4 April 1912.

41. *Ibid.*, 22 September 1912.

42. Gwyneth Logan to Harry Logan, 12 December 1919, LAC.

43. Wolf, *Deliver Me from Pain*, 42. As Joanna Bourke demonstrated in her recent study of medical professionalism and sympathy from the eighteenth century to the twentieth century, male physicians sought to demonstrate in their practice an active, rational, scientific and 'masculine' version of sympathy that was increasingly contrasted with women's perceived emotional sentimentality. See Joanna Bourke, 'Pain, Sympathy and the Medical Encounter between the Mid-Eighteenth and the Mid-Twentieth Centuries', *Historical Research*, 85(229) (2012): 430–52.

44. Elisabeth Robinson Scovil noted that 'the pains are very severe' and anaesthetic relief was all the more required for first-time mothers: *Preparation for Motherhood* (Philadelphia: Henry Altemus, 1896), 265.

45. Wendy Mitchinson, *Giving Birth in Canada, 1900–1950* (University of Toronto Press, 2002), 162.

46. *Selected Journals of L.M. Montgomery*, 22 September 1912.

47. Lucy Bending, *The Representation of Bodily Pain in Late-Nineteenth Century English Culture* (Oxford: Clarendon Press, 2000), 124.

48. Wolf, *Deliver Me from Pain*, 17.

49. Diary of Lucy Ronalds Harris, 28 February 1868, *The Eldon House Diaries*.

50. Diary of Amelia Ryerse Harris, 5 May 1860, *The Eldon House Diaries*.
51. *The Selected Journals of L.M. Montgomery*, 30 June 1912.
52. Though Kirkby was writing from rural California, her experiences would have likely struck a chord with those of many rural Canadian homesteading women. Diary of Georgina Bruce Kirkby, 3 February 1853, in Erna Oilafson Hellerstein, Leslie Parker Hume and Karen M. Offen (eds), *Victorian Women: A Documentary Account of Women's Lives in Nineteenth-Century England, France and the United States* (Stanford University Press, 1981), 213.
53. Mary Kough Brown Journals, 24 January 1865, Brown-Hendry Collection, Hamilton Public Library.
54. Diary of Frances Tweedie Milne, 10 September 1870, AO.
55. Gwyneth Logan to Harry Logan, 17 November 1919, LAC.
56. Walzer Leavitt and Walton, '"Down to Death's Door"', 160.
57. Jo Oppenheimer has demonstrated that in Ontario, the majority of births took place in the home until 1938. As Ontario was, arguably, the most 'modern' and medicalised of Canada's provinces, these statistics are particularly striking. It is important to keep in mind, however, that physicians were increasingly attending deliveries within the home, particularly after provincial legislation effectively criminalised midwifery. Jo Oppenheimer, 'Childbirth in Ontario: The Transition from Home to Hospital in the Early Twentieth Century', *Ontario History*, 75 (1983): 36–60.
58. Diary of Eliza Jane Wilson, 20 June 1904, Glenbow.
59. Diary of Eliza Jane Wilson, 28 June 1904, Glenbow.
60. Mitchinson, *Giving Birth in Canada*, 7.
61. *Ibid.*, 301–2.
62. The medical discourse of the period consistently asserted that because of their luxurious and unnatural lifestyles and 'evolved' sensitivity, white, well-to-do, urban-dwelling women, referred to by one physician as 'the luxurious daughters of artificial life', would experience the highest levels of pain in giving birth. M.L. Holbrook, *Parturition without Pain: A Code of Directions for Escaping from the Primal Curse* (Toronto: Maclear and Company, 1875), 17, 20.
63. Mary Kough Brown Journals, 29 January 1865, HPL.
64. Evelyn Cartier Springett, *For My Children's Children* (Montreal: Unity Press, 1937), 97–98.
65. *Selected Journals of L.M. Montgomery*, 22 September 1912.
66. Agnes Just Read, *Letters of Long Ago* (Caldwell, ID: Caxton Printers, 1936), 25.
67. Diary of Jennie Curran, January 1877, LAC.
68. After taking off her wedding ring due to swollen hands (a common complaint), Harris commented that she 'had an idea perhaps that I might never put it on again' and expressed her anxiety at the fact that she had not yet made a will. Diary of Lucy Ronalds Harris, 11 May 1870 and 24 March 1880, *The Eldon House Diaries*.
69. *Selected Journals of L.M. Montgomery*, 24 May 1914, 148.
70. Physicians consistently focused on 'female troubles, so disastrous to the happiness of women' in the prescriptive literature. Through advice manuals, like Dr Ira Warren's 1884 *Household Physician*, women were repeatedly told that they were 'subject to a class of distressing complaints peculiar to themselves, involving considerations of a delicate nature'. See Mary R. Melendy, *Ladies Home Companion: A Book Giving Full Information on all the Mysterious*

and Complex Matters Pertaining to Women (Brantford: Bradley Garretson Company, 1903), 17; and Ira Warren, *Warren's Household Physician* (Boston: Ira Bradley and Company, 1884), 339. For more on women's relationships with physicians during this period, see Wendy Mitchinson, *The Nature of their Bodies: Women and their Doctors in Victorian Canada* (University of Toronto Press, 1991), 223.

71. The 1896 *Ladies Book of Useful Information* that prescribed the balm also asserted that, aside from 'troublesome cramps of the legs' and 'palpitation of the heart', 'pregnant women are generally affected with heartburn, sickness of a morning, headache, and that troublesome disease, toothache, which accompanies pregnancy'. *The Ladies' Book of Useful Information: Compiled From Many Sources* (London, ON: London Printing and Lithographing Company, 1896), 125–8.

72. Diary of Jennie Curran, 13 February 1879, LAC.

73. Diary of Lucy Ronalds Harris, entries 7 May 1868, 21 February 1877 and 14 May 1880, *The Eldon House Diaries*.

74. Gwyneth Logan to Harry Logan, 31 October 1919 and 17 November 1919, LAC.

75. Dr Pye Henry Chavasse, for example, asserted that 'sickness frequently comes on in the beginning of labour and may continue during the whole process ... sickness in labour is a rather favourable symptom and ... does good'. Pye Henry Chavasse, *Advice to a Wife on the Management of Her Own Health, And on the Treatment of Some of the Complaints Incidental to Pregnancy, Labour, and Suckling, with an Introductory Chapter Especially Addressed to a Young Wife* (Toronto: Hunter Rose Company, 1879), 203–4.

76. Diary of Frances Tweedie Milne, 8 October 1870, AO.

77. Diary of Constance Kerr Sissons, 12 January 1903, edited by Rosalind Kerr, in *The Small Details of Life*, 22.

78. Diary of Eliza Jane Wilson, 30 June 1904, Glenbow.

79. *Selected Journals of L.M. Montgomery*, 22 September 1912.

80. Wolf, *Deliver Me from Pain*, 6.

81. Mitchinson, *The Nature of their Bodies*, 193. Indeed, as Judith Walzer Leavitt has shown, women did not always passively 'go along' with this process, but at times actively agitated for medical interventions, including anaesthesia. Judith Walzer Leavitt, 'Birthing and Anesthesia: The Debate over Twilight Sleep', *Signs*, 6(1) (1980): 147–64.

82. Nancy M. Theriot, 'Women's Voices in Nineteenth-Century Medical Discourse: A Step Toward Deconstructing Science', *Signs*, 19(1) (1993): 2, emphasis in original.

I am very grateful to Rob Boddice, Esme Cleall, Orisheweyimi Olugbo and Laura Schwartz for their helpful suggestions on how best to improve this chapter.

12
'The Agony of Despair': Pain and the Cultural Script of Infanticide in England and Wales, 1860–1960

Daniel J.R. Grey

Pain – both physical and emotional – was central to the cultural script of infanticide in England and Wales between 1860 and 1960. Despite Elaine Scarry's famous suggestion that 'Whatever pain achieves, it achieves in part through its unsharability',[1] recent scholarship has emphasised that pain is indeed both 'shareable' through visual, textual and oral means, and that it is assigned historically and culturally specific meanings by those who witness and experience it.[2] By 'cultural script', in this instance, I refer to a commonly accepted set of assumptions about what it meant to be 'infanticidal' between the late nineteenth and mid-twentieth centuries, beliefs that were shared by a diverse cohort of English and Welsh men and women. This included, but was not confined to, politicians from across the political spectrum, doctors, lawyers and social scientists. It was generally understood that the 'typical' infanticide defendant was a young woman who had been perceived as virtuous and hard-working prior to the crime, was then seduced and abandoned by a feckless or malicious man under false promises of marriage and 'respectable motherhood', and was left facing economic and reputational ruin if her condition were discovered. The more closely a woman could 'fit' the story of her individual circumstances into this narrative, the more likely it was that a judge and jury would view her sympathetically.[3] The physical agony of childbirth was an essential part of any discussion of infanticide in this period, as was the accompanying fear and shame that would surround an illicit and possibly secret pregnancy and birth.[4] While such ideas were certainly not uncontested both within and between different groups, it is a striking feature of the discourses that surrounded the subject of infanticide in England and Wales that they were, for the most part, marked more by their similarities than their differences. Critics of the judicial treatment of infanticide diverged between the minority, who felt that its distressing nature called for especially severe punishment, and the majority, who advocated more sympathetic responses that took account of a defendant's social and psychological circumstances. Yet these opposing factions were united in their depiction of the 'typical' case and the belief that it was a 'special' crime that deserved particular consideration.

This chapter explores how and why the closely entwined experiences of pain, shame and fear became – and remained – key elements of the construction of infanticide in England and Wales for at least a century.[5] The passing of the Infanticide Act 1922 and the amending Act of 1938, which allowed women who killed their babies while suffering from 'mental disturbance' to be treated as if they were guilty of manslaughter rather than murder, was widely believed to have introduced a substantially more compassionate and fair way of dealing with such defendants than if they faced the capital charge.[6] Both before and after these laws were passed, however, lay and professional discussions of infanticide continued to stress a key theme, namely that it was the combination of physical pain and mental anguish that was primarily to blame for the crime, and that the vast majority of defendants therefore should not be held fully responsible for their actions. Indeed, the notion that the pain and trauma of giving birth could trigger temporary and murderous psychosis in otherwise normal individuals meant that this cultural script could be extended to married women who killed their children. Until now, however, historians have largely ignored the issue of pain in the depiction of infanticide, despite (or perhaps because of) its ubiquity in both medico-legal and cultural accounts. This chapter examines how the narrative of pain was drawn upon through an examination of infanticide cases that were tried at the assizes, using surviving witness statements taken before a coroner or magistrate as a way of locating relevant trials.[7] As the following case illustrates, both medical and lay testimony in these cases was routinely expected to interpret potential signs of physical pain and emotional distress on the part of the defendant when giving evidence.

In April 1864, Jane Lewis was sweeping the yard at her home in West Bromwich when she was interrupted by a cry from next door: 'Mrs. Lewis ... for God's sake come here I am very ill.'[8] Hurrying to her neighbour's house, she found 22-year-old Elizabeth Taylor crouching, pale and sweating, and trying vainly to mop up a small amount blood spattered across her kitchen floor. Concerned and exasperated in equal measure, Lewis asked why she had not been called earlier and insisted that a doctor, and preferably an additional neighbour, should be sent for to help her. Taylor vehemently denied any need for medical attention, adding, as Lewis later recounted, 'that there was no neighbour that liked her and she wished no one besides me'.[9] As Taylor's condition continued to deteriorate, Mrs Lewis decided that a doctor had to be sent for urgently. The local physician arrived and after conducting a brief examination, he surprisingly announced to Mrs Lewis and Taylor that she must be 'fomented with Cold'.[10] Indeed, when Taylor remained confined to the kitchen bench with little improvement over the course of the day, he was willing to prescribe fever medicines and provide a certificate that evening which decreed 'that Elizabeth Taylor was suffering from haemorrhage of the womb brought on probably by cold'.[11] If this sounds preposterous to twenty-first-century ears, it is worth noting that it sounded equally so to the inhabitants of the village – particularly so some hours after the doctor's visit, when Jane Lewis, who had previously helped four local women besides herself give birth, delivered what she immediately recognised as a placenta during a particularly severe bout of abdominal pain. It seems that Lewis assumed at this point that Taylor had either had a miscarriage or stillbirth, and gently pressed her neighbour for more details of what had happened while Taylor cried bitterly and insisted there had been no child. When Lewis stepped into the coal cellar to fetch more fuel for the kitchen fire, however, such comforting impressions collapsed as she caught sight of a white object near the bottom of the steps: 'I called out For God's sake Betsy what have you done and I wept bitterly as I found it was the body of a child. She cried and hid her head in the pillow and made no answer.'[12]

Taylor was tried for the murder of her baby at the Staffordshire Summer Assizes in July 1864.[13] William Kite, the surgeon who performed the post-mortem, had already testified before the coroner that the child had almost certainly breathed and – even more importantly – that he believed the baby had possessed what was referred to in medical jurisprudence as a 'separate existence' from its mother.[14] Determining 'separate existence' was an essential part of any trial for infant homicide between 1860 and 1960. It relied on conducting a battery of tests during the autopsy that were potentially inconclusive. If the results did not fit precisely into the complicated rules and case law that governed these matters, it was impossible to sustain a charge of murder or manslaughter.[15] Kite reiterated this opinion when he testified at the assizes and concluded that the cause of death had been multiple fractures of the skull, possibly hastened by an attempt at suffocation. Despite this statement and the evidence of the other witnesses, as with the overwhelming majority of infanticide defendants in nineteenth-century England

and Wales, Taylor was convicted of the lesser offence of concealment of birth rather than of either murder or manslaughter.[16] The resulting sentence of 12 months' hard labour was certainly not a minor punishment, especially in comparison with the end of the period under discussion. By the 1950s, the assize courts invariably issued probation orders to those convicted of infanticide and focused on providing offenders with psychiatric treatment. Yet it was also, of course, much less severe than the potential sentence of death. The last woman to be executed in England and Wales for the murder of her own baby was Rebecca Smith in 1849, but proposals to amend the law and formally separate cases of newborn murder from other sorts of homicide remained deeply controversial right through to the interwar years.[17] As such, the narratives of pain, shame and distress remained essential components of the 'story' of infanticide in England and Wales, a way to explain a heinous and shocking crime by a figure who was nonetheless often viewed with a great deal of sympathy.[18] Two broad themes emerge: pain of the body and emotional pain.

Pain of the body

In many cases of suspected infanticide during the late nineteenth and early twentieth centuries, there was a pressing need for the authorities to try and determine whether or not an accused woman might actually have *known* that she was pregnant before she found herself in the throes of childbirth, quite apart from the possibility that she might (consciously or otherwise) have denied such knowledge to herself as well as to others.[19] This was not a trivial question, particularly since a significant number of accused women claimed that the onset of labour had taken them by surprise and successfully argued at their trial that this physical shock and their unpreparedness – even in cases, such as the 1918 trial of Dora Lewenstein, where it was not a first pregnancy[20] – had either contributed to or actively caused the death of their child. Even during the mid-twentieth century, the provision of formal sex education in Britain was a highly controversial and anxiety-inducing subject with regard to both the factual content covered and how this material was delivered.[21] For the most part, this meant that knowledge about sexuality and reproduction (including contraception) was usually gained and disseminated in a haphazard manner through the general population.[22] Since access to this information was further mediated by constraints of class and gender, the majority of working-class British women only learned about these matters (not always passed on clearly or accurately) from family or neighbours when they married, and could retain significant gaps in their knowledge until the birth of their first child was imminent.[23] Despite the fact that she already had a three-year-old daughter and was cohabiting with the girl's father when she secretly gave birth to and killed her second child in September 1943, 22-year-old Elsie Dutton later told the medical officer at Manchester Prison that she had not realised she was pregnant again, even though her periods had stopped and she had noticeably put on weight.[24] Moreover, even for qualified physicians, the diagnosis of pregnancy before 'quickening' – the moment the foetus begins to move – could be a complicated matter and it was acknowledged that symptoms such as bodily aches, cessation of the menses, vomiting and conspicuous weight gain might well signify a woman's illness and were not *necessarily* indications of impending motherhood.[25]

With these factors in mind, it is sometimes difficult to tell from the depositions whether or not women later accused of newborn child murder who denied their pregnancy might have been genuinely ignorant of and frightened by their situation, or at least uncertain of it prior to the onset of labour pains, rather than making a last-ditch attempt to ward off discovery by neighbours or employers. Mary Ellen Moore, a 23-year-old waitress from Liverpool, had asked the advice of both her stepmother and of her landlady, Martha Rimmer, as to what she should do after being frightened by the sudden cessation of her menses in early 1910, and consulted more than one doctor in an effort to solve this problem.[26] Despite Rimmer's blunt assessment, delivered after hearing her lodger complain repeatedly of nausea and backaches, and seeing Moore without her corsets in the kitchen, 'that if she had been married I should have thought she was 6 or 7 months pregnant', Moore vehemently denied this possibility and seems to have been genuinely and vocally outraged when the nurse and doctor she later saw at Shaw Street Women's Hospital concurred with her landlady's opinion.[27]

Moore was by no means unusual among infanticide defendants in thinking of what was actually her pregnancy or even the onset of labour itself as symptoms of severe illness and pain, or in describing it as such to those around her. Complaints of headaches, cramps, backaches, leg pain and 'flooding' (haemorrhage), as well as specific illnesses such as influenza or food poisoning, were frequently used by infanticide defendants between 1860 and 1960 in order to explain their sudden debility to friends and colleagues, and aimed to allay any possible suspicions that they were 'in the family way'.[28] Mary Ann Osborn, a 25-year-old farm worker from the Welsh village of Cadoxton, was found prostrate in the stables by a fellow servant, Mary Sutton, on the latter's return from milking in April 1881. Osborn was visibly unwell and claimed to have been 'Over the fields ... very sick and very poorly' since the early hours of the morning, when Sutton had heard her go downstairs to the privy more than once.[29] When her condition did not improve after an hour sat by the fire with sweet tea, it was decided that Osborn needed proper rest and nursing to recuperate from her sudden affliction, and her employer drove her by cart to her mother's house several miles away. It was not until Osborn was already being taken home that Mary Sutton, noticing spots of blood and a freshly washed floor, discovered the corpse of a baby boy hidden under wooden sticks and nightsoil in the privy.[30] Women might also attempt to pass off the physical mess that had been caused by giving birth as the direct result of other bodily ailments: Fanny Gane argued vociferously with a hotel manageress in 1891 that the bloodstains that were found splattered across towels, carpet and bedclothes in her room on the Isle of Wight had resulted from a profuse nosebleed rather than a secret miscarriage, as the other woman suspected.[31] Unsurprisingly, this desperate strategy was attempted only by those for whom other efforts to hide the signs of their delivery, such as soiled blankets or clothing, had already proved impossible. For many women, the appearance of sudden and inexplicable bloodstains (however small) or a noticeably slimmer figure was the immediate precursor to their arrest and trial at the assizes. Mary McDonnell, a 19-year-old Irish woman working for a middle-class couple in Nottingham, had been questioned repeatedly by two of her fellow servants as to whether or not she was pregnant. This was a charge that she had always staunchly denied. However, when the housekeeper noticed a suspicious stain on McDonnell's overalls in November 1951, it prompted her to immediately search the younger woman's bedroom, where she discovered the body of a child wrapped in a sheet and hidden in the wardrobe.[32]

Conversely, witnesses might also be anxious to demonstrate to the courts that they had initially accepted this 'illness' and discomfort at face value, and that there had been no reason to think otherwise given the prior good character of the accused. The day before she gave birth to (and most likely killed) her infant son, 23-year-old Margaret Hopkins had been planting potatoes at the County Durham farm where she worked. Too unwell to return to the fields the next morning, Hopkins told her employers that she was suddenly struck down by an unspecified sickness, and her mistress was keen to emphasise when giving evidence to the coroner in May 1876 that there had been 'nothing in her appearance that might lead me to suspect that she was in the family way'.[33] No suspicion that this 'illness' and its symptoms might actually have concealed an illicit birth and the disposal of a corpse seem to have been raised until the news circulated that a dead baby had been discovered in a field nearby. Even after this, the question of Hopkins' involvement remained uncertain until a local physician confirmed that she had recently had a child.[34]

Bearing in mind Jacqueline Wolf's observation that women who have given birth explain their experiences of parturition very differently,[35] it is interesting that there seems to have never been any question, even amongst the most unsympathetic commentators on infanticide, that the accused woman had suffered great pain around the time the crime was supposed to have occurred. In this sense, the construction of the infanticidal woman departed significantly from the claimed insensitivity to physical pain (and, indeed, emotional distress) that was commonly ascribed to other sorts of violent offender by criminologists in the late nineteenth centuy, as part of a broader attempt to delineate clear boundaries between supposedly 'normal' and 'abnormal' individuals at this time.[36] Instead, her physical pain as she went through the agony of childbirth acted to *reinscribe* her very normality and the connection of the infanticidal woman with the rest of society, placing a strong emphasis on her previous 'respectable' character and adherence to appropriate standards of femininity, rather than marking her out as deviant. Given the medical consensus in both Europe and North America by the 1890s that middle- and upper-class women were unable to stand the agony of childbirth unaided,[37] this sensitivity was in itself a tangible demonstration of the infanticidal woman's previous good character. However, perhaps the most significant factor that ultimately determined how an accused woman was treated by the criminal justice system was the perception of her individual circumstances and her emotional state at the time of the crime.

Emotional pain

In some infanticide cases, clear evidence of mental illness on the part of the accused provided a relatively easy explanation of the crime for the jury and provoked a good deal of sympathy for her plight. The belief that the intense pain of childbirth might, quite literally, drive women mad can be traced back to at least as far as the Middle Ages in Britain and Europe, if not before.[38] Such ideas – albeit in modified form – remained in circulation during the nineteenth century, despite the fact that the advent of anaesthesia offered a potential (though sometimes controversial) solution to women suffering through parturition.[39] Maria Curry, the 34-year-old wife of a stevedore, was universally considered by both medical and lay witnesses during her trial in February 1901 at the Central Criminal Court to have been suffering from 'puerperal insanity', a condition broadly analogous to present-day postnatal depression, when she had drowned her three-month-old son and then attempted to commit suicide.[40] Repeatedly described as a sober, industrious and loving mother to the dead baby and her four older children, and married to a respectable man, Curry was noted to have been 'queer in her mind the last fortnight'.[41] During this time, she had eaten very little, suffered from insomnia, seemed to be anxious and distressed, and had complained of frequent pains in her arms and head, as well as resorting to self-medication with alcohol.

A dangerous condition which might strike vulnerable women at any time between the onset of pregnancy and after breastfeeding had begun, puerperal insanity could display a variety of symptoms including and beyond those that were exhibited by Curry. But it was widely accepted during the nineteenth century that there was a high risk of sufferers attempting infanticide and/or suicide, even if few such women actually carried out these acts.[42] When combined with the numerous descriptions of her as having been a 'good mother' and happily married before falling ill, it is therefore unsurprising that Curry was promptly found 'guilty but insane' by the jury and sentenced to be kept in custody as a criminal lunatic at Her Majesty's Pleasure.[43] The popular association of child homicide with temporary mental illness – even when this did not formally correspond to the definitions laid down in tests determining insanity and criminal responsibility– was sufficiently strong that a very high proportion of married or widowed women who killed their children were found 'guilty but insane' in the nineteenth century, despite the fact that it was regularly suggested that unmarried mothers were by far the most vulnerable to it.[44] As one doctor put it during a discussion of puerperal insanity and infanticide at the Medico-Legal Society in 1928, it was commonly understood that a formerly respectable woman might well kill her illegitimate child within 24 hours of giving birth as a result of 'the agony of despair'.[45] Even though British psychiatrists were increasingly questioning the validity of puerperal insanity as a condition in its own right during the early twentieth century, its ongoing *cultural* usefulness as an explanation for why a woman might commit infanticide meant that it was not only implicitly written into the wording of the 1922 Infanticide Act, but 'insanity of the puerperium' continued to be cited as a distinct category in textbooks written by other branches of the medical profession as late as the Second World War.[46]

Insanity was not, however, a prerequisite for judicial or popular sympathy in cases of infanticide. When Fanny Gane was tried for the murder of her newborn son at the Winchester Assizes in 1891, there was no suggestion on the part of the defence that she had been insane at the time.[47] Investigations had also revealed that Gane was already the mother of a six-year-old girl who was living with her own mother.[48] Despite this, and the earlier damning verdict of the coroner who had given as his opinion 'that the prisoner murdered the child, and meant to murder it',[49] she was repeatedly depicted as a tragic figure rather than a malevolent one. When the jury returned their verdict of 'guilty of murder', Gane collapsed, and after being revived began to scream, while the judge placed the black cap on his head and pronounced sentence.[50] She continued to scream while she was being removed from the court and begged of anyone in earshot 'Don't tell my mother'.[51] Describing Gane's case as being of 'a painful nature', the *Hampshire Advertiser* reassured its readers that 'the recommendation of the jury for leniency will have considerable weight in this case'.[52] Although the sentence was quickly commuted to penal servitude for life and the Home Office agreed when reviewing her file that 'there was nothing ... to take the case out of the usual category',[53] the substantial sympathy that seems to have been felt for Gane in Hampshire meant that there was a small but well-organised campaign for her release that continued to run for at least five years after her conviction.

In other cases, it was suggested with varying degrees of explicitness that the crime might have been avoided altogether if the father of the unfortunate child had been willing to marry the accused woman.[54] Whether or not love and romance alone provided a sufficiently stable foundation for a successful marriage was increasingly questioned in the wake of the First World War,[55] but in those cases where pre-marital sexual activity had led to pregnancy, the expectation in most working-class communities was generally that the man should put any qualms aside, accept his responsibility for the situation and solve it by going through with a hastily arranged marriage to his sweetheart. Ena Garwood told the detective who interviewed her at Brighton in 1931 that the killing of her newborn son was ultimately the fault of her boyfriend, Norman Wyatt, for refusing to marry her when she became pregnant: 'if he had stuck to me through it, like I asked him to, I should never have done what I did'.[56] Rather than proposing as Garwood had evidently expected, Wyatt instead procured some sort of abortifacient which she refused to take, presumably in the hope that the closer she came to giving birth, the greater the likelihood that he would change his mind.[57] When the baby was born, Garwood claimed to have been in so much pain that she killed him without really knowing what she was doing, although her confession remained more resentful of Wyatt's 'unmanly' conduct that compounded her shame than it was remorseful. She concluded bitterly that his horror and fury when she later admitted to him that she had killed the child showed 'that is all he cares'.[58] Given her statement under caution and the opinion of the doctor who performed the post-mortem that the baby had indeed had a 'separate existence' and died from strangulation with a ligature, Garwood seems to have been lucky that the grand jury took a sympathetic view of her circumstances as a desperate single woman and threw out the bill for murder.[59] Having confessed to the secondary charge of infanticide at the Lewes Summer Assizes, she was released on her own recognisances for the sum of £5.

Freda James, a 22-year-old housekeeper who confessed to infanticide at the Chester Summer Assizes in 1960, gave a relatively full statement to police, a practice that only seems to have become common in the records of those cases surviving at the National Archives that were compiled during the 1950s. Her confession stated that she had begun to suffer haemorrhaging and then passed out. When she regained awareness, she had been in a nightmarish state when, as the statement recorded:

I was terrified in case I had left the poor thing alive. I did not know what I was doing it was horrible. I tied the stocking round its neck in case it was suffering and then put it in a case in the wardrobe. I didn't know what I was doing and I can't remember much, it was horrible.[60]

Essential to James' explanation for the crime, however, was the overwhelming sense of isolation and despair that she felt at her circumstances. Until about a year earlier, she had spent months pining for her fiancé before she was forced to accept that a longed-for letter providing her with a ticket to New Zealand to join him and finally marry was never going to arrive. Her new boyfriend, the father of her unfortunate baby, had also guessed she was pregnant after she was taken violently ill during one of their meetings. He had immediately severed all contact with her. Shortly after the discovery of her child's body, she became violently ill and was admitted to hospital, drifting in and out of consciousness and experiencing auditory hallucinations as well as vaginal bleeding. Although her mental and physical condition had improved markedly by the time of her trial, her psychiatrist noted that 'if she does not further improve the question of readmission to hospital may well arise'.[61] The court agreed with his assessment and James was given a three-year probation order that allowed – indeed, insisted on – her continued attendance as an outpatient at the local psychiatric hospital.

Conclusions

When examined together, the cases of infanticide I have discussed in this chapter show distinct similarities in their representation and their outcomes, despite the fact that they were tried at different assize courts and in different decades. Although they include women in a wide range of personal situations, the pain of the body and emotional distress were subjects that routinely appeared in the evidence of witnesses. While descriptions of physical and emotional pain can be elusive and some-times fragmented in this testimony, reflecting the fact that 'depositions are complex documents with no clear authorial voice',[62] these issues nonetheless remained central to the attempts made by a range of com-mentators fully to understand the crime of infanticide and ultimately to determine the most fair and appropriate mechanisms for dealing with it. Most defendants had attempted to deflect suspicion by claim-ing that the bodily changes brought on by pregnancy and the agony of childbirth were in fact symptoms of acute illness, such as Elsie Dutton saying to her partner that 'I have had my monthlies, I have got a pain in my leg and my back'.[63] In many cases, the crime was discovered because women who were trying to conceal their condition often went into labour in spaces where they had very limited privacy, being forced to explain why they were not at work or performing household tasks, and were then equally constrained in their ability either to hide the evidence of their delivery or to dispose of the unfortunate child's body.

Perhaps most important were the attempts to explain the reason for their crime, which invariably rested on the shame and anxiety single mothers faced in this period. Mary McDonnell admitted to police that once she realised she was pregnant, her overriding concern was that: 'I didn't want to lose my job. The reason I told the untruths was because I was ashamed and wanted to keep the trouble to myself.'[64] Although married women had very different concerns and no need to fear the stigma of illegitimacy, they were still at risk of puerperal insanity, including its most severe and violent forms. Since all women who gave birth were held to be potentially at risk from the 'mental disturbance' later defined by the Infanticide Acts, this often provided juries with an answer for a crime that seemed to otherwise defy explanation. A regular feature of depositions was that the accused women frequently claimed to have committed the act when the pain of childbirth left them in a profoundly disoriented state. Even where this did not meet the formal tests of insanity under the law, such explanations could garner a great deal of sympathy from the jury when taken in conjunction with the recently lost respectability that was associated with most infanticide defendants. These traits meant that there were practical as well as moral reasons to treat these women sympathetically: some commentators recorded their view that women who were convicted of infanticide were uniquely 'reclaimable' among the ranks of female offenders. In assessing the suitability of Mary Ellen Moore for transfer to a rescue home after five years' imprisonment, the Duchess of Bedford noted approvingly in 1914 that 'so many of these [infanticide] cases have passed out to respectable marriage or honourable service'.[65]

There is of course a danger in drawing too close a comparison between cases of infanticide committed by women across the course of a century and from different regions of England and Wales. A potential risk in examining the accounts of physical and emotional pain in infanticide cases from across England and Wales between 1860 and 1960 is to elide the specificities of these cases and the broader shifts in culture and society that occurred during this period, not least in terms of gender relations, sexual knowledge and attitudes to single motherhood.[66] Nor, of course, does this script of women's bodily and emotional pain fit in any way with the different yet equally contested discourses which surrounded the killing of infants by their fathers, and it is important to point out that men seem consistently to have accounted throughout this period for a significant minority of those accused of infanticide.[67] However, examining these depositions helps to illuminate the significant degree of continuity in the discourses of infanticide in England and Wales, features that resonate even now with the twenty-first-century cases that continue to be dealt with under the Infanticide Act 1938.

Notes

1. Elaine Scarry, *The Body in Pain: The Making and Unmaking of the World* (Oxford University Press, 1985), 4.
2. See, for example, Roselyne Rey, *The History of Pain* (trans. Louise Elliott Wallace) (Cambridge, MA: Harvard University Press, 1995); Lucy Bending, *The Representation of Bodily Pain in Late Nineteenth-Century English Culture* (Oxford: Clarendon Press, 2000); Esther Cohen, *The Modulated Scream: Pain*

in Late Medieval Culture (University of Chicago Press, 2009); Javier Moscoso, *Pain: A Cultural History* (trans. Sarah Thomas and Paul House) (Basingstoke: Palgrave Macmillan, 2012); the contributions to 'Perspectives on Pain', *19: Interdisciplinary Studies in the Long Nineteenth Century*, 15 (2012), www.19.bbk. ac.uk/index.php/19/issue/view/85 (date accessed 24 February 2014); and the collected essays in this volume.

3. Daniel J.R. Grey, '"Almost Unknown Amongst the Jews": Jewish Women and Infanticide in London 1890–1918', *London Journal*, 37(2) (2012): 122–35.

4. Margaret L. Arnot, 'Gender in Focus: Infanticide in England, 1840–1880' (PhD thesis, University of Essex, 1994); Daniel J.R. Grey, 'Discourses of Infanticide in England, 1880–1922' (PhD thesis, Roehampton University, 2008). This association of pain and shame also seems to have been a common theme in discussions of infanticide in Ireland and Scotland. See Elaine Farrell, *'A Most Diabolical Deed': Infanticide and Irish Society, 1850–1900* (Manchester University Press, 2013); Anne-Marie Kilday, *A History of Infanticide in Britain c. 1600 to the Present* (Basingstoke: Palgrave Macmillan, 2013); Clíona Rattigan, *What Else Could I Do?: Single Mothers and Infanticide, Ireland 1900–1950* (Dublin: Irish Academic Press, 2011).

5. These ideas represented a reworking of early modern understandings of the relationship between shame, pain and infanticide: see Kilday, *A History of Infanticide*, 51–76.

6. Daniel J.R. Grey, 'Women's Policy Networks and the Infanticide Act 1922', *Twentieth Century British History*, 21(4) (2010): 441–63.

7. Depositions held at The National Archives (TNA) represent a sample of pre-trial testimony to the coroner or magistrate that was preserved by the clerks of the assize courts. Even though infanticide is better represented in this respect than other serious crimes, there remain very substantial gaps in the records for this period. Moreover, all such archival material includes a number of files that remain closed for the foreseeable future due to concerns about data protection.

8. TNA ASSI 6/10. *R v. Taylor.* Deposition of Jane Lewis.

9. *Ibid.*

10. *Ibid.*

11. According to Mrs Lewis, Taylor's father had directly requested this certificate, since his other adult children were suspicious that Elizabeth either was or had recently been pregnant.

12. *Ibid.*

13. TNA ASSI 2/39. See 23 July 1864 (unpaginated).

14. *Birmingham Daily Post*, 4 May 1864, 3.

15. For this reason, the case notes of the eminent pathologist Sir Bernard Spilsbury always included reference to whether or not he had been able to determine 'separate existence' in the 146 metropolitan post-mortems he conducted where newborn murder was suspected: see Wellcome Library PP/SPI. Papers of Sir Bernard Spilsbury, 1905–45. See also the discussion of this complex issue in Alfred Swaine Taylor, *The Principles and Practice of Medical Jurisprudence*, 2 vols, 2nd edn (London: John Churchill and Sons, 1865), 884–988; John Glaister and Edgar Rentoul, *Medical Jurisprudence and Toxicology*, 10th edn (Edinburgh: E. & S. Livingstone, 1957), 372–95.

16. *Birmingham Daily Post*, 25 July 1864, 3.

17. Grey, 'Women's Policy Networks'. On the case of Rebecca Smith, see Katherine D. Watson, 'Religion, Community and the Infanticidal Mother: Evidence from 1840s Rural Wiltshire', *Family & Community History*, 11(2) (2008): 116–33.
18. Arnot, 'Gender in Focus'; Grey, 'Discourses'; Kilday, *A History of Infanticide*.
19. On the question of denying pregnancy even to oneself, see Margaret L. Arnot, 'Understanding Women Committing Newborn Child Murder in Victorian England', in Shani D'Cruze (ed.), *Everyday Violence in Britain, 1850–1950: Gender and Class* (London: Longman, 2000), 55–69.
20. TNA CRIM 1/174/3. Deposition under caution of Dora Lewenstein. See also Grey, '"Almost Unknown Amongst the Jews"', 131–2.
21. Lesley A. Hall, 'In Ignorance and in Knowledge: Reflections on the History of Sex Education in Britain', in Lutz D.H. Sauerteig and Roger Davidson (eds), *Shaping Sexual Knowledge: A Cultural History of Sex Education in Twentieth Century Europe* (Abingdon: Routledge, 2009), 19–36; Hera Cook, 'Getting "Foolishly Hot and Bothered"? Parents and Teachers and Sex Education in the 1940s', *Sex Education*, 12(5) (2012): 555–67.
22. Hera Cook, *The Long Sexual Revolution: English Women, Sex and Contraception 1800–1975* (Oxford University Press, 2004).
23. *Ibid.*; Lucinda McCray Beier, '"We were Green as Grass": Learning about Sex and Reproduction in Three Working-Class Lancashire Communities, 1900–1970', *Social History of Medicine*, 16(3) (2003): 461–80.
24. TNA ASSI 65/47/1. *R. v. Dutton*. See letter to the Clerk of Chester Assizes from E.W.A. Cormack, medical officer at HMP Manchester, 30 October 1943. After she confessed to infanticide at her trial, Dutton was released on her own recognisance for £5, with an additional requirement that she must remain at the local Salvation Army Home 'so long as Major Buck thinks necessary'. See TNA ASSI 64/166. *R. v. Dutton*.
25. See Graily Hewitt, *The Pathology, Diagnosis and Treatment of Diseases of Women: Including Diagnosis of Pregnancy*, 3rd edn (London: Longman, Greens & Co., 1872); W.S.A. Griffith, 'A Clinical Lecture on Diagnosis of Pregnancy', *British Medical Journal*, (1903): 833; J.M. Robson, 'Pregnancy Diagnosis in Theory and Practice', *British Medical Journal*, (1934): 1063.
26. TNA ASSI 52/160. *R. v. Moore*. Deposition of Martha Rimmer.
27. *Ibid.* Moore was convicted for the murder of her newborn daughter at the Liverpool Summer Assizes a few months after this incident, although the sentence of death was promptly commuted to penal servitude for life. See TNA ASSI 51/117. *R. v. Moore*; *The Times*, 21 July 1910, 4.
28. This was also common practice by accused women in earlier decades: see Samantha Williams, 'The Experience of Pregnancy and Childbirth for Unmarried Mothers in London, 1760–1866', *Women's History Review*, 20(1) (2011): 67–86.
29. TNA ASSI 72/3. *R. v. Osborn*. Deposition of Mary Sutton.
30. *Ibid.* Osborn was found guilty of concealing the birth of her child at the Swansea Spring Assizes and sentenced to two months' hard labour. See *Western Mail*, 16 May 1881, 3.
31. TNA ASSI 26/27. *R. v. Gane*. Deposition of Louisa Mary Long.
32. TNA ASSI 13/232. *R. v. McDonnell*. Deposition of Annie Gertrude Leigh. After she confessed to infanticide at her trial in February 1952, McDonnell was issued with a 12-month Probation Order.

33. TNA DURH 18/2. *R. v. Hopkins*. Deposition of Margaret O'Brien. Hopkins was later acquitted on all charges: TNA DURH 15/10. See 5 July 1876 (unpaginated); *York Herald*, 6 July 1876, 3.

34. TNA DURH 18/2. *R. v. Hopkins*. Deposition of Hugh Clark. Given the controversy surrounding medical examinations of infanticide defendants at this time, it is interesting that Clark was very careful to explain he had obtained Hopkins' consent to this procedure. See Daniel J.R. Grey, '"What Woman is Safe...?": Coerced Medical Examinations, Suspected Infanticide, and the Response of the Women's Movement in Britain, 1871–1881', *Women's History Review*, 22(3) (2013): 403–21.

35. Jacqueline H. Wolf, *Deliver Me from Pain: Anesthesia and Birth in America* (Baltimore: Johns Hopkins University Press, 2009), 1–12.

36. Bending, *The Representation of Bodily Pain*, 207–224.

37. Wolf, *Deliver Me from Pain*, 48. See also Whitney Wood, Chapter 11, this volume.

38. Cohen, *The Modulated Scream*, 137.

39. Moscoso, *Pain: A Cultural History*, 96–104.

40. TNA CRIM 1/65/3; *Old Bailey Proceedings Online* (www.oldbaileyonline.org, version 7.0, date accessed 24 February 2014), February 1901, trial of MARIE [*sic*] CURRY (34) (t19010204-172).

41. TNA CRIM 1/65/3. Deposition of James Curry.

42. Hilary Marland, *Dangerous Motherhood: Insanity and Childbirth in Victorian Britain* (Basingstoke: Palgrave Macmillan, 2004), 167–200; Kilday, *A History of Infanticide*, 169–181.

43. *The Times*, 7 February 1901, 13.

44. A recent study of female child-murderers incarcerated at Broadmoor Asylum between 1863 and 1884 found that 83 per cent of the 150 women in this sample were either married or widowed at the time of their committal. See Alison Pedley, '"A Painful Case of a Woman in a Temporary Fit of Insanity": A Study of Women Admitted to Broadmoor Criminal Lunatic Asylum between 1863 and 1884 for the Murder of their Children' (MA thesis, University of Roehampton, 2012), 23.

45. *The Lancet*, 3 March 1928, 447.

46. Aleck W. Bourne, *Synopsis of Obstetrics and Gynaecology*, 9th edn (Bristol: John Wright and Sons, 1945), 224–6.

47. *Hampshire Advertiser*, 12 December 1891, 8.

48. TNA ASSI 26/27. *R. v. Gane*. Deposition of William Pugsley.

49. *Richmond & Twickenham Times*, 29 August 1891, 9. Although Gane had killed her infant on the Isle of Wight, she lived and worked as a general servant in Mortlake and was forced to return there without having been able to dispose of the body beforehand, hence the coroner's inquest was held in Surrey.

50. *Hampshire Advertiser*, 12 December 1891, 8.

51. *Ibid.*

52. *Ibid.*

53. TNA HO 144/240/A53169. Civil service note on file, 20 August 1892.

54. Complaints that the fathers of murdered children shared at least some of the blame for the crime by abandoning the mother had a long history: see Grey, 'What Woman is Safe...?'.

55. Claire Langhamer, *The English in Love: The Intimate Story of an Emotional Revolution* (Oxford University Press, 2013).
56. TNA ASSI 36/44. *R. v. Garwood.* Deposition of Thomas Wells.
57. On the involvement of male partners in actual or attempted abortions, see Emma L. Jones, 'Abortion in England 1861–1967' (PhD thesis, University of London, 2007), 81–119.
58. TNA ASSI 36/44. *R. v. Garwood.* Deposition of Thomas Wells.
59. TNA ASSI 31/58. See 7 July 1931 (unpaginated); East Sussex Record Office COR 3/2/1931/25.
60. TNA ASSI 84/277. *R. v. James.* Statement under caution of Freda Moreen James.
61. *Ibid.* Report of Dr K.C.S. Edwards, Consultant Psychiatrist at Deva Hospital, Chester, 28 May 1960.
62. Louise A. Jackson, *Child Sexual Abuse in Victorian England* (London: Routledge, 2000), 29.
63. TNA ASSI 65/47/1. *R. v. Dutton.* Deposition of Bernard Maguire.
64. TNA ASSI 13/232. *R. v. McDonnell.* Statement under caution of Mary McDonnell.
65. TNA HO 144/1092/195705. Letter to Charles Dryhurst from Adeline, Duchess of Bedford, 13 August 1914.
66. Lesley A. Hall, *Sex, Gender and Social Change in Britain since 1880*, 2nd edn (Basingstoke: Palgrave Macmillan, 2013).
67. On the differing treatment of men accused of child homicide, see Ginger Frost, '"I am Master Here": Illegitimacy, Masculinity and Violence in Victorian England', in Lucy Delap, Ben Griffin and Abigail Wills (eds), *The Politics of Domestic Authority in Britain since 1800* (Basingstoke: Palgrave Macmillan, 2009), 27–42; Elaine F. Farrell, '"The Fellow Said it was Not Harm and Only Tricks": The Role of the Father in Suspected Cases of Infanticide in Ireland, 1850–1900', *Journal of Social History*, 45(4) (2012): 990–1004; Jade Shepherd, '"One of the Best Fathers Until He Went Out of His Mind": Paternal Child-Murder, 1864–1900', *Journal of Victorian Culture*, 18(1) (2013): 17–35.

13
Imagining Another's Pain: Privilege and Limitation in Parent and Child Relations

Linda Raphael

Theorising the Other

What do we talk about when we talk about knowing another's pain? One way of answering this question is to turn to fiction. In this turn, I suggest we can attain a greater understanding of the privileges and limitations involved in knowing another's pain. When we read fiction, we have access to characters' inner lives in a way that gives form to feeling. We have opportunities to 'witness' the effects of interactions with others on various individuals, and in our role as narrative audience, we frequently experience affects similar to those of the characters. In other words, fictional representations involve us cognitively and emotively in human experience, much of which inevitably involves 'hurt feelings'.

The two stories analysed in this chapter, Lorrie Moore's 'People Like That are the Only People Here: Canonical Babbling in Peed Onk'[1] and Jhumpa Lahiri's 'Unaccustomed Earth',[2] feature a parent and child in situations that involve knowing another's pain. My choice of each story for this discussion is determined by the intricate ways that language and face-to-face encounters sometimes facilitate and other times impair characters knowing the pain of the other. In the rich narratives of each story, characters attempt to imagine the other's pain and are alternately abetted and hindered by the closeness of their relationship. Clearly, the desire to imagine the pain of the other, including others who are or have been part of the lives of the parent and child, informs each story. Yet the stories involve differences that are significant for this study: one represents painful emotions arising from a sudden, unanticipated and critical event. The pressures exerted by the emergent nature of the situation in 'People Like That' offer a glimpse into the effects of frightening and unexpected events on emotions and relationships. 'Unaccustomed

220

Earth', on the other hand, spans, through memory, the lifetime of the characters and gives voice to emotions 'recollected in tranquility'. The similarity of the stories – each one involves a relationship, parent and child, that typically arouses intense emotions – gives a symmetry to the two stories, while the pain represented in each has enough difference to broaden the scope of the study.

The general question of what it is to know another – in particular, to know another person's pain – has been a subject of philosophers since the time of the Ancient Greeks. The historian and philosopher Javier Moscoso, in an explanation of the trickiness of identifying with others' pain, contends that the quest to sympathise with the other involves a 'shuffling back and forth' from the 'immediate sensations' of one person and the 'mediated sensations' of another that makes it impossible to distinguish between emotions and judgement. He analyses the lack of differentiation between feeling and judging in the following manner:

> The boundary between reality and fiction becomes very unclear, and further still: utterly diffuse. On the one hand we can sympathize with imaginary beings. On the other, we can also ignore the suffering of real victims under the cover of their situation's dramatic qualities. Real pain can seem fictitious to us, and fictitious suffering may seem real.[3]

The problem Moscoso describes is similar to the one Rousseau expressed concerning actors and theatregoers expressing sympathy at dramatic performances, without any sense of how their sympathies might be significant for real-world situations. Furthermore, Moscoso's statement gives a twist to David Hume's belief that ethics are based on feelings rather than abstract moral principles,[4] and Adam Smith's 'impartial spectator', the term for a dialectical engagement of the self with an imaginary other as the other being watched by oneself.[5] In other words, rather than taking Moscoso's claim to disparage the nature of sympathetic engagement in general, we can read in it an appreciation of fiction as a way of knowing the other, as well as a general scepticism about our ability to know the mind of the other.

Among the twentieth-century theories on knowing other minds, Ludwig Wittgenstein's 'ordinary language' philosophy challenges the sceptical argument that language is inadequate to the challenge of knowing another by acknowledging that we cannot know precisely what pain feels like to the other, nor can we know what she sees when she calls an object 'red' or any other colour, but we can communicate about pain, and have feelings about the other's pain, just as we can dependably attribute a specific colour to an object.[6] Following Wittgenstein, Stanley Cavell emphasises the significance of acknowledging the impossibility of fully knowing the other, even as one attempts to understand her, as a sign of respect for the other.[7] In a similar vein, Emmanuel Levinas' belief that one's subjectivity inheres in the recognition of the alterity of the other, expressed in the face-to-face encounter, offers a way to consider the essence of the other as irreducible to oneself.[8] These theories might well lead one to ask what happens to the self in this focus on sympathising with the other; one significant response comes in the work of Bernard Williams, who centres on ethics and a personalist philosophy that incorporates Hume's position on feelings, and adds to these other theories concerning self and other a recognition of the significance of one's own projects and desires to his life.[9]

Each of these philosophical theories takes exception to some aspects of the others; however, they are all concerned with offering ways to think about the need for any social being to care about the pain of the other, to attempt to know this pain in some personally felt way, to live with the difficulty of that challenge and to respect the other as a sign of one's own subjective self.[10] That the focus of theories of knowing the other is knowing the other's pain is due to the fact that any group or society depends for its welfare on the ability of its members to recognise and to care about the suffering of the other.

While close analysis of a fictional text offers a unique appreciation of what it is to know another's pain in terms of these philosophical theories, readers typically engage not only cognitively but also affectively with a text. Ideally, the author and the reader inhabit the perspectives of the characters in the text; at the same time, they are aware of this relationship as one of alterity. Michael Marais puts the matter succinctly in a discussion of ethics and engagement in the fiction of J.M. Coetzee:

> The point of Blanchot's meditation on the impossibility of the writer's desire for the Other, that is, the impossibility of either actualizing or eliminating radical otherness, is that the literary text *always* describes the excess of an involvement with an alterity which it cannot totalize. This description is not reducible to reference or representation: it is a function of the interruption of the text's totality, its *inevitable* failure to instantiate or negate the excess of the Other. In consequence, the space of the literary text is, for Blanchot, always an approach of the Other: every work dissembles an Orphic descent.[11]

Similarly, Judith Butler contends that the autonomy of a fictional character is central to an ethical reading of fiction. Citing the example of Catherine Morland in Henry James' *Washington Square*, she reasons that Catherine's motives at the conclusion of the novella (to refuse the suitor whom she had previously desired and who had 'jilted' her) are not named. She is, according to Butler, the exemplary autonomous fictional character.[12] I would add that Catherine's act has an affective resonance, perhaps of pain, perhaps of satisfaction, for the reader, despite (or perhaps in part because of) a lack of access to her motives.

The representation of emotions in fiction, whether they are directly expressed or left to the reader's imaginative construction, accounts in large part for a reader's connection with characters as well as for a greater understanding of various emotions. In *Pride, Shame, and Guilt: Emotions of Self-Assessment*, the philosopher Gabrielle Taylor takes up examples from fiction to elaborate on her skilful analysis of emotions. For example, in James Joyce's *The Dead*, the protagonist Gabriel, who has just delivered a speech to a gathering of people at his aunts' home, finds himself in an elevated mood following the event and feels particularly fond and close to his wife, Greta, on the way home. In this openness to feeling, the couple's conversation leads to an intimate confession that Greta had once been in love with a boy from Galway. Gabriel makes an ironic comment that perhaps that is why she had wanted to make a visit to Galway, to which she replies that the boy is dead; she tells Gabriel that she believes that he died for her because he left a sickbed to see her before she left the city. Gabriel's mood changes in a flash:

> Gabriel felt humiliated by the failure of his irony and by the evocation of this figure from the dead, a boy in the gasworks. While he had been full of memories of their secret life together, full of tenderness and joy and desire, she had been comparing him in her mind with another. A shameful consciousness of his own person assailed him. He saw himself as a ludicrous figure, acting as a penny boy for his aunts, a nervous, well-meaning sentimentalist, orating to Vulgarians and idealising his own clownish lusts, the pitiable fatuous fellow he had caught a glimpse of in the mirror. Instinctively he turned his back more to the light lest she might see the shame that burned upon his forehead.[13]

The power of the lonely and profound emotion of humiliation, realised in the extraordinary language of Joyce, has affective consequences for the reader, who, as Moscoso says, is as likely to sympathise with pain through a fictional representation as any real-life experience of pain.

The ubiquity of pain in a social environment is dramatised in a humorously astute passage in Dickens' *Hard Times*. Mrs Gradgrind responds to her daughter Louisa's query '"Are you in pain, dear Mother?"': after having noted that it is 'something new when anyone wants to hear of [her]', she responds: '"I think there's a pain somewhere in the room, but I could not positively say that I have got it".'[14] At the same time, Mrs Gradgrind's strange retort implies that pain may be experienced as a response to its existence 'somewhere', even if the pain did not originate in one's own body or mind. Some fictional works depend almost entirely on a reader's experience of characters' pain; for example, in Kazuo Ishiguro's *The Remains of the Day*, the protagonist Stevens, a butler, is 'painfully' unaware of his motivations and his feelings. Without the reader's experience of the pain associated with Stevens' actions and omissions, the novel would be ethically and emotionally unrewarding. At the same time, the 'unknowability' of Stevens remains central to the novel, as the disagreements of critics over his desires and ethics attest.[15]

'People Like That are the Only People Here: Canonical Babbling in Peed Onk'

In this story a one-year-old boy is diagnosed with a cancerous tumour of the kidney (Wilms' tumour). When the Mother wonders about the Baby's pain, the Surgeon tells her: 'The baby won't suffer as much as you.'[16] Imagining what the Baby will experience, she thinks:

Who can say what babies do with their agony and shock? Not they themselves. They put it all no place anyone can really see. They are like a different race, a different species: they seem not to experience pain the way we do. Yeah, that's it: their nervous systems are not as fully formed, and they just don't experience what we do. A tune to keep humming through the war.[17]

The Mother is a writer, 'words are her trade'. Moore emphasises the way that language is used, or misused, throughout the story. A more subtle expression of communication in the story involves face-to-face encounters, or the lack thereof.

The 'war' begins in the radiology department: the naked Baby and Mother are physically engaged, as she holds him on the table. The Radiologist holds a 'cold scanning disk'. Twice the anxious Mother asks if he is finding something, and twice the Radiologist responds that the surgeon will speak to her. The Mother makes another attempt for his sympathy and reassurance by telling him that her uncle had a kidney removed for something that turned out to be benign. The Radiologist deflects her transparent plea with a 'broad, ominous smile. "That's always the way it is," he says. "You don't know exactly what it is until it's in the bucket," and with a grin [that] grows scarily wider' offers the confused Mother the explanation that his expression is 'doctor talk':

'It's very appealing... It's a very appealing way to talk', the Mother responds. She imagines swirls of bile and blood, mustard and maroon in a pail the colors of an African flag or some exuberant salad bar: *in the bucket* – she imagines it all.

The Radiologist's parting casual gesture, a tousling of the Baby's 'ring-letty hair' as he comments 'Cute kid', communicates no more sympathy for the Mother's anguished state than a feigned compassion that seems to be part of an off-hand repertoire of language and gesture.[18] Pamela Schaff and Johanna Shapiro term the discourse of the doctors in the story: 'concrete clichés of detachment, juxtaposed against the Mother's vigorous, literate, grammatical attempts at integrating the facts and feelings of her son's diagnosis [that] provide a stark and painful repre-sentation of professional narrative's limitations'.[19] The Mother, in this first encounter with her child's illness, vacillates between the expression and regulation of painful emotions. In a study of emotional regula-tion, several psychologists conclude: 'Emotions are powerful, elusive, dynamic processes. They have the capacity to regulate other processes and to be regulated.'[20] While this assertion should surprise no one who has given thought to the way emotions work, the study is a reminder of the dialogic effect of emotions in any human encounter.

The Mother and Baby's meeting with the Surgeon begins no more auspiciously than the one with the Radiologist. Replete with representa-tions of emotional regulation, the encounter is continually interrupted: 'He has stepped in, then stepped out, and then come back in again. He has crisp, frowning features.' 'The Mother knows that her own face is a big white dumpling of worry', but the Surgeon's quick movements in and out of the room preclude his recognition of her face. The Baby, however, is engaged with his Mother's face as he pulls her hood to cover it. In a more startling form of peek-a-boo, the Baby plays with the light switch, alternately creating dark and light so that when the Surgeon pronounces that the Baby has a Wilms' tumour, the Surgeon is 'suddenly plunged into darkness'. The lights go on and off, and the Mother, at a loss in this metaphorical and real darkness, resorts to the exactitude of language to make contact with the Surgeon: 'Is that apostrophe "s" or "s" apostrophe?' she asks. The narrator refers to the Mother's profession as a writer and teacher for an ironic explanation of the offbeat query that is at the same time an attempt to establish a small but significant connection with the Surgeon and to convey and/or to hide her emotions.[21]

The Mother attempts to make the tumour familiar, something that belongs in her lexicon rather than the unknown and unwelcome stranger that it is. The word belongs to the Surgeon, not to her; she thinks 'She has never been at a time like this before'.[22] Perhaps she and the Surgeon could come face to face over this term that belongs to him, not to her, but now so strangely to the Baby. The Surgeon might perform as Anatole Broyard, writer and editor, wished his surgeon would:

> I would like my doctor to understand that beneath my surface cheerfulness, I feel what Ernest Becker called 'the panic inherent in creation' and 'the suction of infinity' [and to] enter my condition and look around at it from the inside like a kind of landlord, with a tenant, trying to see how he could make the premises more livable.[23]

But the Radiologist does not 'enter her condition'. He responds:

> 'S apostrophe... I think'; the lights go back out, but the Surgeon continues speaking in the dark. 'A malignant tumor on the left kidney... We will start with an article nephrectomy', says the Surgeon, instantly thrown into darkness again. His voice comes from nowhere and everywhere at once.

His following words – that they will begin with chemotherapy after that – fail to shed any brightness in the room. With their faces invisible to one another, the Mother says to the Surgeon: '"I've never heard of a baby having chemo"... Baby and Chemo, she thinks: they should never even appear in the same sentence together.' In the darkness, listening to words that are foreign to her, the Mother feels more keenly the injustice (as if there is justice in illness) of the Baby having a tumour.[24]

Lacking a sympathetic relation with the doctors, the Mother retreats inward, exploring her repertoire of beliefs and common superstitions for explanations of the unwonted diagnosis:

> Wait a minute. Hold on here. The Baby is only a baby, fed on organic applesauce and soy milk – a little prince! – and he *was* standing so

close to her during the ultrasound. How could he have this terrible thing? It must have been her kidney. A fifties kidney. A DDT kidney. The Mother clears her throat. 'Is it possible it was my kidney on the scan? I mean, I've never heard of a baby with a tumor, and, frankly, I was standing very, close.' She would make the blood hers, the tumor hers; it would all be some treacherous, farcical mistake.[25]

But the Surgeon misses the call for help in her technically flawed suggestion: 'No, that's not possible', he responds. His language is exact (even though medicine is not) and as the lights go off again, the Mother, Baby and Surgeon are all invisible and incomprehensible to one another. With either an insight into maternal guilt or a tendency toward clichés, the Surgeon says, as quoted above, 'The Baby won't suffer as much as you':

> The Mother thinks, 'Who can contradict?' Not the Baby, who in his Slavic Betty Boop voice can say only mama, dada, cheese, ice, bye-bye, outside, boogie-boogie, goody-good, eddy-eddy, and car. (Who is Eddy? They have no idea.) This will not suffice to express his mortal suffering.

Ironically, the Baby's language represents the early desire to communicate with others, while the Surgeon's apparent flippancy underscores his desire to avoid language that denotes the fear and/or other emotions that a parent would be experiencing. He tells the Mother that she 'will get through it' and responds to her question 'How' by saying: 'You just put your head down and go.' The Mother thinks of him as a 'skilled manual laborer. The tricky emotional stuff is not to his liking'.[26] If she follows his advice, she will avoid face-to-face encounters by assuming the posture and attitude of a football player. In *Illness Narratives*, the medical anthropologist Arthur Kleinman imagines that physicians who relate to 'the layers of meaning in a patient's narrative are not too different from the complex layers that constitute the "thickness of surface" with "infinitely receding depths" that have been attributed to Sophoclean tragedy'.[27] This tragedy is what the Mother is fearing; it is what the physicians hope to avoid, both by effective medical treatment and by avoiding a recognition of the 'complex layers' that underlie the Mother's questions.

At the same time, similarly suffering individuals express a complicated relation to the Mother's 'Sophoclean tragedy', as the story's title anticipates. Beginning with the Mother and the Husband, Moore represents the difficulty of responding to others who are suffering similarly. Like the 'impartial spectator' figured by Adam Smith, the fellow sufferer may bring to an encounter with the other less recognition of the other's alterity and more sense of the self as the prime sufferer. The Husband, who shares the Mother's anxiety, advises her to write – they will need the money for hospital bills. Following a sardonic scene during which the Mother and Husband rant at fate and entreat an imaginary 'Manager' of chance, 'the Husband buries his face in his hands'.[28] The Mother comes face to face not with the Husband, but with herself:

> Her face, when she glimpses it in a mirror, is cold and bloated with shock, her eyes scarlet and shrunk. She has already started to wear sunglasses indoors. Like a celebrity widow. From where will her strength come? From some philosophy? From some frigid little philosophy?[29]

Turned in towards the self, the sufferer is unable to fully relate to another. Instead, intense suffering may result in a search for one's responsibility for the crisis that causes the pain as a way of attempting to rationalise the suffering. The Mother takes the well-worn path to maternal guilt to rationalise the inexplicable horror of her child's cancer. Her ludicrous guilty imaginings resonate with a reader on two counts: first, historical images of the ideal mother are introduced by the Mother's thought that 'all of life has led her here' and expanded in the list of her 'transgressions'; and, second, the literary technique of using irony to include the reader in the community of those who 'get it' is inclusive largely due to representations of mothering throughout history.[30] In antiquity, female virtue was associated with motherhood; a poem from the Augustan period that 'comes from the grave': 'Yet I lived long enough to earn the matron's robe of honour, nor was I snatched away from a childless house ... all my children came to my funeral.'[31] The poem suggests that if a woman's child dies before she does, she has not lived an honourable life.

The Mother imagines that 'her occasional desire to kiss the Baby passionately on the mouth (to make out with her baby!)' constitutes the transgression for which the Baby's tumour is retribution.[32] Her fear resembles a particular sort of maternal angst that 'fanned the flames of fears of dishonor' in Ancient Greece concerning a woman's breast – that it is not only nurturing, but can be incestuous. The bare breast, and especially lactation, was sometimes regarded as threatening, leading to fears that 'the boundary between the motherly and the erotic might be blurred'.[33]

Moore's wild-seeming creation of maternal angst resonates with the historical record of women interpreting aspects of pregnancy and childbirth. Historically, mothering has been associated with pain in terms of childbirth (physical pain to the mother and/or child and fears of loss of one or both of them), and emotional vulnerability according to superstitions and fits, and with the history of societal and self-blame of mothers for their children's troubles.[34] Included in the litany is the ordinary matter of leaving the Baby with babysitters, recalling John Bowlby, who believed, as the historian Angela Davis puts it, that 'the mere physical separation from the mother was a pathogenic factor in a child's development'.[35] As the Baby flicks off the lights once more, the Mother begins to cry, thinking: 'All of life has led her here, to this moment. After this there is no more life.'

Moving from more serious transgressions to the ridiculous, the Mother recalls 'placing a bowl of Cheerios on the floor for him, as if he is a dog', making jokes about mothering and about baby talk, and chastises herself for believing in alternative medicine, which she now regards as 'the wacko maiden aunt to chemo'. She imagines that 'Now her Baby, for all these reasons – lack of motherly gratitude, motherly attention' has cancer.[36] The power of the ironic, even ludic, imaginings resides in its reminder to the reader of the social world the Mother inhabits and to implicitly critique this world from her point of view. As one critic points out: 'Lorrie Moore, like Slavoj Zizek, like[s] jokes. We are freed from the burden of acting "as if" and can momentarily acknowledge to each other that we are all in on the masquerade of this half-real life.'[37]

The next section begins with 'Take notes' and is followed by: 'In the end, you suffer alone. But at the beginning you suffer with a whole lot of others.'[38] In her first encounter with another parent, the Mother is scolded when the Baby grabs hold of a deflated rubber ball that lies close to four-year-old Ned. Unaware that the ball attaches to a tube that draws fluid from Ned's liver, the Mother simply tells the Baby to 'share'; Ned's mother shouts at the Baby, who then cries. Either the Baby's cries or her anxiety over Ned – or both – cause Ned's mother to cry. Ned and the Mother, the two who are not crying, look at one another, and Ned responds to the Mother, who says that she is 'stupid' for having thought that the rubber at the end of the tube was a ball. Ned agrees, not with the idea of her stupidity, as his mother seems to, but with her innocent mistaking of an object associated with a child to be a ball and not a tube attached to his liver: 'It does look like a toy', and he smiles.[39] The interchange distinguishes between the language and face-to-face encounter of the experienced mother and the neophyte in this unwanted place. At the same time, the words and the face (smile) of the four-year-old child convey a sympathy that is made possible by a lack of cognitive sophistication and consequent fearlessness about entering into the other's place.

While Ned's mother's reaction to the Mother implies that the newcomer is not yet appropriately initiated, the other parents share stories that are intended to include the novice into the collection of suffering parents. Here there is 'pain somewhere in the room', as Mrs Gradgrind avers. In the Tiny Tim Lounge, the Mother hears stories of childhood: 'leukemia in kindergarten, sarcomas in Little League, neuroblatomas discovered at summer camp'. She thinks that 'there is a kind of bravery in the air that isn't brave at all. It is automatic, unflinching, a mix of man and machine'. A father says to her: 'Everyone admires you for your courage. They have no idea what they are talking about.' When he says this, the Mother thinks: 'I could get out of here... I could just get on a bus and go, never come back.'[40] Indeed, what does bravery or courage have to do with one's response to a child's cancer? Synonyms for the words include 'nerve', pluck', 'valor', 'daring' and 'audacity'; the opposite is 'cowardice'. If one does not respond 'courageously' to a cancer that first showed itself at Little League, is she cowardly? The father's candour appeals to the Mother, and the content of his message reminds her of how deeply this strange world of childhood cancer, with its 'canonical babbling', is alien to her.

Compared to 'canonical babbling', the Mother reasons: 'How can it be described? How can any of it be described? The trip and the story of the trip are always two different things... One cannot go to no place and speak of it; one cannot both see and say, not really.'[41] The cynical rhetoric belies the significance of the narration. The affective power of the words on the reader may well be to intensify the meaning of the words. The language that obscures the pain nonetheless communicates it by saying: 'It is not this way; imagine how it is, Reader!' The narrative creates a longing for words that tell, and in the generative tension between the narrative and the reader, the reader needs to work hard – to look between the lines, to supply meaning when it does not readily come. Like all marital language, the narrator claims that 'overheard or recorded, all marital conversation sounds as if someone must be joking, though usually no one is'.[42] The language of this narrative is not joking.

The potential power of language and face-to-face encounters come together in the relationship between the Mother and Frank, the father who criticised the use of the word 'courage'. Despite the estranging stories of the parents in the lounge, the Mother yearns for stories, as does the reader who stays with this story. Frank offers 'I could tell you stories' and the Mother responds 'Tell me one. Tell me the worst one'. He answers: 'The worst? They are all the worst. Here's one: one morning I went out for breakfast with my buddy – it was the only time I'd left Joey alone ever; left him for two hours is all – and when I came back, his N-G tube was full of blood. They had the suction on too high and it was sucking the guts right out of him.' The Mother understands; the same thing had happened 'to us', she replies.⁴³

While the Mother shares this particular experience with Frank, she must realise his otherness in the terms of his story: his son has been treated for cancer for four years. 'She feels flooded with affection and mourning for this man.' Joey's mother fills in the spot opened for the 'bad mother' in the Mother's imagined litany of self-blaming acts: she left Joey's father, and to a great extent Joey, two years after his cancer diagnosis, married again and has another child. She visits only occasionally. Frank, on the other hand, 'shaved his head bald in solidarity' with Joey, an act that recognises the significance of the face-to-face encounter.⁴⁴ He and Joey face one another and in their same baldness, they recognise the other, while Joey's alterity is unchallenged because he will leave Frank. Frank's exchange with the Mother on the subject of 'the worst story' does not partake of language of false hope, amelioration, either of a physical or emotional sort. The words they use together are not sympathetic platitudes; they are words that convey desire for the worst news (so that one can look at it and can say 'It cannot be worse than that'). These words describe but do not analyse their pain. The limits of language are one subject of the story – the privilege of knowing the other through language that does not attempt to do what cannot be done, at least through direct language; the other subject, the face-to-face encounter, as opposed to the earlier darkness, is made once by Ned's innocent acknowledgment of the Mother's bewilderment, and finally in Frank's bald acknowledgment of his son's autonomy – Joey's looming death that defines the absolute nature of otherness, even or especially in the case of deep attachment to the other.

'Unaccustomed Earth'

In the opening scene of this story, a grown daughter imagines her father's loneliness following the sudden death of her mother. The father has been sending the daughter postcards while he has been travelling, a form of communication that is at once revealing (open to any viewer) and concealing (brief and often impersonal). The father, Dadu's, brief visit to the daughter, Ruma, and her young son, Akash, who have recently moved from Brooklyn to Seattle due to Ruma's husband, Adam's, job, figure as the narrative present of the story. The sections of the story are distinguished by representations of the daughter's and father's consciousness, a technique that enables Lahiri to express the discrepancies between the daughter's imaginings of her Bengali-born-and-raised father and the father's (in part Americanised) view of matters, as well as the differences between Ruma's and her father's thoughts about her.

At the same time, in this third-person narration, Lahiri takes advantage of the effects of free indirect discourse to develop the intersubjective nature of the father/daughter relationship. In many instances, an idea, fact or feeling is reported without its being attributed to the characters' consciousness; instead, the report indicates the mood of the visit. Free indirect discourse has the advantage of suggesting a place where we might imagine the face-to-face encounter with the other, where knowing of the other does not depend on the character whose consciousness is represented by that language being fully aware of the reported thought or feeling.[45]

This complex way of knowing the other imitates a psychological reality during which one's encounter with the other involves intuitions of thoughts and feelings that may or may not be apparent to the other. Such is the recognition of the other – an acknowledgment of the separateness of self and other at the same time as one engages imaginatively with the other. For example, the narrator gives a factual account about Adam's absence during the visit: 'Adam would be away that week, on another business trip. He worked for a hedge fund, and had yet to spend two consecutive weeks at home since the move. Tagging along with him wasn't an option. He never went anywhere interesting.'[46] 'Another' trip suggests the tedium of a partner's frequent absences from home; 'had yet' confers a sense of longing; 'tagging' suggests Ruma's feeling of dependence (she had been a high-paid lawyer until just after her mother's recent death and her move). However, the degree to which Adam's travel troubles Ruma is undetermined; further, the reader does not know whether she is consciously thinking about this at the narrative time when it is reported. Over and again, such reports create the mood of the story and render the characters' lives more fully, without determining the conscious connection the characters have to some of the elements of their lives at given moments. As in 'People Like That', the author does not, in Blanchot's words, overcome the otherness of the characters.

The memories of the mother that bind Dadu and Ruma, even though they do not speak of her, resonate with Lahiri's ability to reflect what one critic refers to as a 'cosmopolitan' identity, rather than a binary opposition between new and old.[47] Dadu thinks of his wife as having 'liv[ed] for these visits to her family', while Ruma remembers her mother visiting when Akash was born, 'sitting alone knitting a blanket for him' and relieving Ruma when she needed sleep by 'holding the baby, refusing to put him down, cradling him in her arms for hours'.[48] She remembers that she kept only three of her mother's 218 saris. The flashes to times with the mother, in both their minds, hints at some things typical of the lives of Bengali women in the following description from a traditional Bengali husband in the 1970s:

My wife passed away with most of her wishes fulfilled. She saw her daughters married and well-established in life. Her only son had a good college career, and she had selected his bride, and her son got his special training in America, and was in a good post in a technical and industrial firm. She was quite happy and proud of her position in life, with a husband and a son upon whom she doted and on whom she could also rely, and with five daughters who were happily married, and her five sons-in-law who were just like her own sons, and her daughters' children who were her loving grandchildren.[49]

The influence of the implied values in the Bengali culture on Ruma's mother, and on Ruma herself, subtly affects the narrative. The Bengali tradition, which, though altered in ways that are in keeping with the changing roles of women, still has resonance in Bengali life. The opaqueness of the influence of the parents' early life on the parents and then on Ruma contributes to her alterity – in her relation to others and to the reader.

In the one instance that the word 'pain' occurs in the narrative, the value of parent/child relations forms the subject. The father, Dadu, remembers that his wife was tormented when Ruma and her brother, Rami, 'were enamored of their newfound independence' in college, and that he had been 'pained' as well, though he never admitted it. His tendency not to voice his feelings accounts for his keeping to himself that Ruma had hurt him – he felt:

condemned by her, on his wife's behalf. She and Ruma were allies, and he had endured his daughter's resentment, never telling Ruma his side of things, never saying that his wife had been overly demanding, unwilling to appreciate the life he'd worked hard to provide.[50]

Similarly, Ruma had kept to herself the criticism she felt from Dadu: when he comments on how much gasoline Ruma's SUV consumes, she feels 'the prick of his criticism as she had all her life'.[51] Although her father had not told Ruma his side of things, she had felt the effects of the criticism that most likely emanated from his resentment at her closeness to her mother. One interpretation of the way Ruma and her father have behaved – keeping their feelings to themselves – finds them wanting in intimacy; another possibility is that they have preserved the chance for intimacy by respecting the other's autonomy.

At present, Ruma and her father's decisions not to share their feelings lead to a misapprehension on Ruma's part. Although she believed (rightly) that her father did not love her mother deeply, she nonetheless anticipates a loneliness that he in fact is not experiencing. This is a misinterpretation of his pain; he had been thinking about how free he is of responsibility, travelling with one suitcase, not having to visit relatives in India, as his wife was wont to do. But:

> Ruma feared that her father would become a responsibility, an added demand, continuously present in a way she was no longer used to. It would mean an end to the family she'd created on her own: herself and Adam and Akash, and the second child that would come in January.[52]

What he has not revealed is that on the several European trips he has taken since his wife's death, he has developed an intimate relationship with a woman. The brevity of messages that postcards carry help him to leave out this important aspect of his travels. However, the reader learns about Mrs Bagchi early in the story, so that they 'read' Ruma's father with significantly different information than Ruma does, imitating the way that one often interprets another in terms of information that cannot be shared but that affects the intersubjective relations. In free indirect discourse, the narrator reports that Mrs Bagchi's:

> voice appealed to him most, well-modulated, her words always measured, as if there were only a limited supply of things she was willing to say on any given day. Perhaps, because he expected so little, he was generous with her, attentive in a way he'd never been in his marriage.[53]

This explanation of Mrs Bagchi's appeal to Dadu expresses his prefer-
ence for simplicity and it also indicates one reason why talking to Ruma
about Mrs Bagchi would be potentially hurtful to Ruma – this woman
is very different from her mother. However, Dadu's awareness of these
feelings is not entirely clear, given that free indirect discourse sometimes reports feelings and thoughts that are not completely evident to
the character.

On the one hand, being secretive can prevent hurt to others; on the
other hand, it can preclude face-to-face encounters. When Dadu shares
slides of his trip and Mrs Bagchi appears in one, he quickly moves to
another slide, but Ruma asks who the Indian woman was. 'He was grateful that the room was dark. That his daughter could not see his face.'
The blush of embarrassment is not in one's control, either to cause or
prevent. His secretiveness, like his reticence to tell Ruma at any time
that her 'siding with her mother' was painful for him, and Ruma's
sensitivity to his criticism, limits their intersubjective understanding
of the other's pain – and of their own. When her father attempts to
encourage Ruma to return to practising law, she is unable to appreciate
his concern and to relate to the points he makes about the significance
of her working. The threat of criticism keeps her from recognising that
he thinks that she is 'now alone in this new place, overwhelmed, without friends, caring for a young child, all reminding him, too much, of
the early years of his marriage, the years for which his wife had never
forgiven him' because she had been taken away from her family.[54] He
had assumed that Ruma's life would be different, and the fact that it is
not apparently pains him. The complication of Dadu and Ruma's intersubjective relationship is that each of them would consider the expression of their concerns too painful for the other: the father cannot say
that he worries that his daughter will be unhappy like her mother and
that he always thought her life would be different from her mother's,
in large part because painful emotions were hidden in the past. Ruma
cannot confide her loneliness to her father (the frequent calls to her
many friends in Brooklyn, where she and Adam had lived, have tapered
off, with no replacements).

What Dadu offers, however, is the creative energy that the quotidian, everyday life and language offer. The first instance of his sensitive
attention to the ordinary occurs when Ruma takes her father to the
room where he will stay, and Akash jumps on the bed. When he tells
his grandson not to jump with his shoes on, Ruma chimes in to say
the same thing. But Akash keeps jumping for several seconds. Asked
again to stop, he asks: 'Why?' To this Dadu answers: 'Because I will have

nightmares.' The child stops jumping – he has his first hint of his ability to affect his grandfather's feelings (he would have nightmares) because of Dadu's sensitive handling of an ordinary matter of a child's obedience.

As Akash quickly develops a strong attachment to his grandfather, the story focuses on the everyday, highly significant matter of an adult's relationship with a child. Akash is the one to first notice that Dadu has gone missing early one morning. Ruma, who is consumed with large life problems, imagines possible disasters. Only a little time passes, however, before Dadu returns from the nursery – not the place Ruma at first thinks, where Akash will go soon – but a plant nursery. He works during the remainder of his stay to create a flower and vegetable garden, with great assistance from Akash. Planting the garden has symbolic and real significance: he had carefully planted and tended a garden in his and his wife's home. This is the one way that he misses her – he grew the vegetables and herbs that she preferred for her recipes, and even after the children were gone, the tradition continued, with his wife inviting friends for dinner, explaining to them that all the vegetables had come from his garden, and sending them home with bagsful. He cannot replace this significant aspect of their lives – he is no cook. It was this aspect of their intersubjectivity, their mutual appreciation of the garden, that expresses the value that their relationship, with its resentments and disappointments, nonetheless achieved.

One afternoon Ruma fixes tea and cookies for herself and her father, and they sit in the yard while Akash plays nearby. Ruma experiences a contentment that has been missing in her life recently. Ordinary matters – the view, the weather, the familiar late afternoon tea-time, shared with her father – ease the pain that she has been feeling, pain over the loss of her mother, her move, the distance she feels from Adam at this time, and the slightly hinted-at anxiety over being at home with a second child. She also realises that her father 'has fallen in love' with Akash. When Dadu speculates 'If I lived here, I would sleep out here [on the porch] in the summer', Ruma answers that he could. Their conversation avoids a full recognition of the offer to have him live with Ruma, yet Dadu later thinks about Ruma's comment and concludes that her offer was made because she needs him. The day before his departure, Ruma repeats the offer more fully: 'I know it would be a big move... But it would be good for you, for all of us. By now Ruma was crying. Her father did not step forward to comfort her. He was silent, waiting for the moment to pass.'[55]

Dadu's silence respects his daughter's otherness, although it might feel taciturn to Ruma (and to readers). His analysis of Ruma's need for him to move in causes him pain: 'the part of him that would never cease to be her father felt obligated to accept'.[56] Yet, he wants to resume his independent life. Having come from a culture where children, parents and grandparents live together, he responds to Ruma's invitation with difficulty in terms of his connection to the past, present and future. Knowing Dadu's thoughts, the reader is in a position to respect his autonomous decision; Ruma has to interpret signs she will soon discover. Furthermore, it is not clear that Ruma believes that she wants her father to live with her rather than feeling a responsibility to have him do so. He believes that she wants him to move with her, and the reader may agree with his analysis, especially because of Ruma's tears. Yet, there is no evidence that her ambivalence has been resolved.

Before he leaves, Dadu instructs Ruma on how to care for the garden, realising as he speaks that she is not listening carefully, just as he had expected. Parts of what he would like for her – that she would return to work, that she and Adam would keep the garden, at least for Akash to enjoy – won't be realised. He is 70 and has accepted a great deal about life's limitations. Thinking about Akash, he reflects that he will not be around to know him in his middle age, and he is saddened. He thinks of how Akash will shut Ruma and Adam out, as he had shut out (left) his parents and as his children had left him. The reality of these things is something Ruma knows as well, but it is generally not something older parents discuss with their grown children because it is too painful. Nonetheless, the fact of mortality exists between them, affecting the mood of their encounters at times. Unlike the less usual situation when children are ill, the parents' ageing is as natural and expected as can be, and yet it is a source of some pain. It is acknowledged in all sorts of ways, from children's independence to the birth of new generations in a family. Yet the grown child can only acknowledge but not take the point of view of the parents' old age and closeness to the end of life, and the parent can acknowledge but not enter into the pain of loss the grown child will experience at the loss of him or her. It is a pain that is different on each side and yet is interdependent – the loss of one another.

When Dadu leaves, Ruma finds a postcard, written by her father in Bengali, save for the name and address of the intended recipient, Mrs Bagchi. She feels as she did when the surgeon announced that her mother had died from an anaphylactic response to the anaesthesia – the postcard announces again the death of her mother. And she thinks that her father has fallen in love, as he was not with her mother, but is with Akash. Yet another sign of the otherness of Ruma's parents is signalled when she considers that if she had taken her mother's attempts to teach her the Bengali language seriously, she could read the postcard. This language belongs to her father, now as before, but not to her. When Akash sees the postcard, he grabs for it, screaming that it is his because Dadu had given him similar-looking objects – the cards that fall out of magazines asking the reader to subscribe – to mark the names of the plantings. The story opens with reflections on the postcards that her father sends to Ruma – and it ends with Ruma deciding to place a stamp on this postcard and put it out for the postman. The postcard features what Ruma thinks of as a generic scene – neither the picture it features nor the message she cannot read tell her intimate details about her father; these she learns only from the recipient's name and her memory of the slide he attempted to elide. The postcard marks a hiding in plain sight that characterises Ruma and Dadu's relationship and makes it possible for them to endure the pain each feels and to derive important pleasure nonetheless. Ruma's mother had given her a lasting gift when Akash was born: she had said: '"He is your meat and bones." It had caused Ruma to acknowledge the supernatural in everyday life.' As she remembers this, she contemplates the awe in death – that an individual 'lives, breathes, eats, is full of a million thoughts, takes up space, and then in an instant is gone, invisible'.[57]

Conclusion

Lahiri accomplishes in 'Unaccustomed Earth' a recognition of the intimate separateness of people. The opening lines of John Bayley's memoir of his marriage to Iris Murdoch tell of the couple, once married, becoming 'closer apart'. Bayley acknowledges the need for and pleasures in intimacy that allow for otherness. The acknowledgment of otherness in a close relationship depends on being secure with the other. All the characters in 'People Like That' and 'Unaccustomed Earth' are dealing with critical, anxiety-producing matters that generally cause them to avoid intimate interactions, whether it is the physicians who are conveying troubling and inconclusive news or the other parents who create some distance from the reality of their situations by 'normalising' them in the language of the everyday, as the collection of parents the Mother meets do. At the conclusion of 'People Like That', the Mother leaves the hospital with the Baby, who is likely cured, and hopes never to see 'these people' again. Pain has made her resistant rather than receptive to others' pain, unless it is expressed in a straightforward manner as Frank's is. In 'Unaccustomed Earth', the pain that fills the room belongs to Dadu and to Ruma, separately. They have lost the same person, the mother, but their experience of that loss is different. Dadu knows that he will be lost to Ruma most of the time, to a life in Pennsylvania and to his travel, and finally in death. Yet, as Bernard Williams writes: 'The condition of my existence is such that unless I am propelled forward by the conatus of desire, project and interest, it is unclear why I should go on at all.'[58] Propelled by his own interest, Dadu leaves Ruma and Akash with a garden, thinking they may not tend it, but knowing that whether they do is not for him to decide. I could have labelled the emotions that are represented in the story: pride, guilt, remorse, sadness, fear, love and so on. But these intersubjectively aroused emotions, realised in large part by the memory of the mother and presence of the child, are fluid, sometimes conflicted, more and less penetrating – and they would be significantly reduced by being named.

The characters in both stories frequently react with a 'bundle of emotions'[59] to painful situations. Their attitudes and/or actions result from a complex source, rather than from only anger, guilt, shame, pride or fear. Sometimes, as in 'People Like That', pain causes a 'flight' reaction, such as the Mother expresses when, at the end of the story, the Baby is released from the hospital: 'I never want to see any of these people again.'[60] Her fear, anxiety, guilt, anger, empathy, sympathy and love are associated with the pain she and other parents experience in the face of a child's critical illness. Similarly, but in a quieter fashion, Dadu experiences a bundle of emotions when he contemplates not knowing his grandson when the latter is middle-aged. His pain includes either resignation or acceptance, depending on how one interprets his feelings. In a somewhat ironic manner, one may take him to be nostalgic for his grandson's future life, even though 'nostalgia' generally refers to something someone experienced in the past.[61] Ruma similarly responds to various emotions – surprise, disappointment, uncertainty, loneliness – when she discovers that her father is involved with a woman. Her emotions lead to acceptance born of love and respect for her father, but also sadness for the loss of her mother and the silences between her and Dadu. The sort of intricate understanding of hurt feelings that each of these stories requires leads to an appreciation of the variety of emotions that such feelings may arouse and the possibility that these emotions may lead to positive attitudes and actions, as well as to withdrawal from the painful situation or rejection of others who are either reminders of the pain or presumed causes of it. Regarding pain as a stimulus to heterogeneous emotional responses is one way to avoid misunderstanding the other, despite the considerable limits on our ability to fully imagine the emotional experience of another.

Notes

1. Lorrie Moore, 'People Like That are the Only People Here: Canonical Babbling in Peed Onk', in *Birds of America* (New York: Picador, 1988), 212–50.
2. Jhumpa Lahiri, 'Unaccustomed Earth', in *Unaccustomed Earth* (New York: Alfred Knopf, 2008).
3. Javier Moscoso, *Pain: A Cultural History* (Basingstoke: Palgrave Macmillan, 2012), 59.
4. David Hume, *A Treatise of Human Nature* (Oxford, Clarendon Press, 1978 [1739–1740]), 385–6.
5. D.D. Raphael, *The Impartial Spectator: Adam Smith's Moral Philosophy* (Oxford University Press, 2007).
6. Ludwig Wittgenstein, *Philosophical Investigations* (Oxford, Basil Blackwell, 1953).
7. Stanley Cavell, 'Knowing and Acknowledging', in *Must We Mean What We Say? A Book of Essays* (Cambridge University Press, 1969), 238–66.
8. Emmanuel Levinas, *Totality and Infinity: An Essay on Exteriority* (trans. Alphonso Lingis) (Pittsburgh, PA: Duquesne University Press, 1969), 43ff.
9. Bernard Williams, 'Persons, Character and Morality', in *Moral Luck* (Cambridge University Press, 1981), 18.
10. Dorothy Hale notes that various theorists (Foucault, Agamben, Adorno, Walter Benjamin, Levinas and Derrida) who have made significant contributions to contemporary thinking about the ethical value of literature express a 'heterogeneity of these political influences [that] has coalesced in a surprisingly unified account of literary value'. Dorothy Hale, 'Aesthetics and the New Ethics: Theorizing the Novel in the Twenty-First Century', *Publications of the Modern Language Association*, 124(3) (2009): 896–905 at 899.
11. Michael Marais, '"Little Enough, Less than Little, Nothing": Ethics, Engagement, and the Change in the Fiction of M.M. Coetzee', *Modern Fiction Studies*, 46(1) (2000): 159–82, at 163, emphasis in original.
12. Judith Butler, 'Values of Difficulty', in Jonathan Culler and Kevin Lamb (eds), *Just Being Difficult? Academic Writing in the Public Arena* (Stanford University Press, 2003), 199–215.
13. Quoted in Gabrielle Taylor, *Pride, Shame, and Guilt: Emotions of Self-Assessment* (Oxford: Clarendon Press, 1985), 108.
14. Charles Dickens, *Hard Times* (Oxford University Press, 1998), 185.
15. See my *Narrative Skepticism: Moral Agency and Representations of Consciousness in Fiction* (Newark, NJ: Associated University Presses, 2001), Chapter 5, and a response to my argument in James Phelan and Mary Patricia Martin, 'The Lessons of Weymouth: Homodiegesis, Unreliability, Ethics, and *The Remains of the Day*', in David Herman (ed.), *Narratologies: New Perspectives on Narrative Analysis* (Columbus, OH; Ohio State University Press, 1999), 88–109.
16. I will follow the text's form of capitalising all the nouns that identify characters' status.
17. Moore, 'People Like That', 218.
18. *Ibid.*, 214.
19. Pamela Schaff and Johanna Shapiro, 'The Limits of Narrative and Culture: Reflections on Lorrie Moore's "People Like That are the Only People Here;

Canonical Babbling in Peed Onk"', *Jounral of Medical Humanities*, 27(1) (2006): 1–17 at 5.
20. Pamela M. Cole, Sarah E. Martin and Tracy A. Dennis, 'Emotional Regulation as a Scientific Construct: Methodological Challenges and Directions for Child Development Research', *Child Development*, 74(2) (2004): 317–33.
21. Moore, 'People Like That', 214–15.
22. *Ibid.*, 215.
23. Anatole Broyard (ed.) *Intoxicated by My Illness, and Other Writings on Life and Death* (New York: Fawcett Columbine, 1992), 42–3.
24. Moore, 'People Like That', 215–16.
25. *Ibid.*, 215.
26. *Ibid.*, 217–18.
27. Harold Schweitzer, 'To Give Suffering a Language', *Literature and Medicine*, 14(2) (1995): 210–21.
28. Moore, 'People Like That', 223.
29. *Ibid.*, 219.
30. See Linda Hutcheon, *Irony's Edge: The Theory and Politics of Irony* (London: Routledge, 1994), esp. Chapter 2, 'The Cutting Edge' and Chapter 4, 'Discursive Communities: How Irony Happens'.
31. Hackworth Petersen and Salzman-Mitchell (eds), *Mothering and Motherhood in Ancient Greece and Rome* (Austin: University of Texas Press, 2012), 8.
32. Moore, 'People Like that', 216.
33. Cindy Stears, 'Breastfeeding and the Good Maternal Body', *Gender & Society*, 13(3) (1999): 308–325; and Douglas E. Gerber, 'The Female Breast in Greek Erotic Literature', *Arethusa*, 11 (1978): 203–12, for an interesting discussion of the eroticism of breastfeeding, contemporaneously and in Ancient Greece.
34. Moscoso discusses many of these superstitions in *Pain: A Cultural History*, 96–104.
35. Angela Davis, *Modern Motherhood: Women and Family in England c. 1945–2000* (Manchester University Press, 2012), 118.
36. Moore, 'People Like that', 216–17.
37. Tom Ratekin, 'Fictional Symptoms in Lorrie Moore's "People Like That are the Only People Here"', *International Journal of Zizek Studies*, 1(4) (2007).
38. Moore, 'People Like That', 224.
39. *Ibid.*, 229.
40. *Ibid.*, 230.
41. *Ibid.*, 237.
42. *Ibid.*, 243.
43. *Ibid.*, 243.
44. *Ibid.*, 246.
45. Free indirect discourse has been a subject of narrative theorists since at least the 1970s. The original work on FID is Kate Hamburger, *The Logic of Literature* (trans. Marilynn J. Rose), 2nd edn (Bloomington: Indiana University Press, 1973). See also Dorrit Cohen, *Transparent Minds: Narrative Modes for Representing Consciousness in Fiction* (Princeton University Press, 1978); Shlomith Rimmon-Kenan, *Narrative Fiction* (London: Metheun, 1983).
46. Lahiri, 'Unaccustomed Earth', 5.
47. Elizabeth Jackson, '"Transcending the Politics of Where You're From": Postcolonial Nationality and Cosmopolitanism in Jhumpa Lahiri's *Interpreter*

of *Maladies'*, *ARIEL: A Review of International English Literature*, 43(1) (2012): 109–25 at 142.
48. Lahiri, 'Unaccustomed Earth', 17.
49. Manisha Roy, *Bengali Women* (University of Chicago Press, 1992), 128–9.
50. Lahiri, 'Unaccustomed Earth', 40.
51. *Ibid.*, 13.
52. *Ibid.*, 7.
53. *Ibid.*, 9.
54. *Ibid.*, 40.
55. *Ibid.*, 58.
56. *Ibid.*,.
57. *Ibid.*, 46.
58. Williams, 'Persons, Character and Morality'.
59. I thank the editor, Rob Boddice, for this term.
60. Moore, 'People Like That', 250.
61. Although literally meaning 'home pain' or 'homesickness', it is not unusual for people to express nostalgia for an earlier time. For example, the destruction of Central and Eastern European places during the Second World War sometimes evokes nostalgic reactions from descendants of inhabitants of those places. For an exploration of the historical pathology of nostalgia, see Susan Matt, *Homesickness: An American History* (Oxford University Press, 2011).

14
Observing Pain, Pain in Observing: Collateral Emotions in International Justice

James Burnham Sedgwick[1]

This chapter explores the effect that observing, investigating, recording and judging atrocities had on participants at the International Military Tribunals (IMTs) held in Nuremberg and Tokyo following the Second World War. It suggests that *confronting pain* in post-conflict environments often *causes pain* and shows how such 'collateral emotions' left enduring marks in Nuremberg and Tokyo on both the tribunals and their participants. In so doing, this research highlights the powerful and unexpected ways in which pain (direct physical, indirect psychological or other forms) can manifest itself in personal and historical change. IMT personnel shared similar experiences. The administration of international justice in post-war societies came with inherent emotional, psychological and personal challenges. In court, participants confronted enemy atrocities. Meanwhile, living in cities ravaged by Allied bombs, tribunal participants also faced their own demons. Civilian personnel felt a particular disconnect between their sanitised view of the war and its devastating, dehumanising ground-level realities. In full view of horrendous destruction, tribunal participants contributed to an optimistic period of international organisation imbued with a renewed, idealistic determination to ensure future world peace. The contrast between high ideals and gritty violence proved a life-altering event for many participants. Ultimately, this chapter demonstrates some of the ways in which observing pain transforms into pain itself. Although in no way comparable to the actual experience of mass trauma, facing atrocity in courts and similar settings constructs pain in emotions and in history. Although grounded in literature from many fields, this is an historical study, not a psychological diagnosis, sociological analysis or scientific investigation. It considers the human and historical contingencies of post-conflict and international humanitarian work to suggest – not prove – ways that in certain sets of circumstances, with particular conditions and involving the right group of actors, emotional rupture can manifest as felt, actualised and realised pain.

The Other in agony: the meaning of pain

Is all pain equal? Can it be shared? This fundamental human experience remains as hard to define as it is universal. Physical pain causes emotional pain, but emotional responses can, in turn, manifest physical outcomes. Pain theorist Elaine Scarry argues that 'what is "remembered" in the body is well remembered'. However, fixing pain – and its legacies – in the body neglects scars on the mind, which, as Chinese author Feng Jicai avers: 'Cannot be washed away.'[2] I want to add a further wrinkle by exploring pain as something shared from a distance. It shows how in some cases the trauma and pain of *others* can cause psychosomatic changes in the *self*.

In the world of international justice and mass atrocity, pain is more than a feeling; it is a personal and political force. Scholars of human rights and genocide hold strong views on the authenticity of suffering. Whose pain 'counts' depends on one's perspective. For some, only the direct trauma of atrocity victims matters. Authors such as Primo Levi, Cathy Caruth and Lawrence Langer, for example, stress the unique disruptiveness of wartime mass trauma. The 'unthinkable', 'unprecedented' or 'incomprehensible' nature of such atrocities shocks victims in such a way that memory of the trauma becomes 'latent': unwanted, repressed, ignored or 'forgotten'; distinct and untransferrable.[3] Even when shared atrocities hurt 'mass' groups of people, scholars like Elaine Scarry contend that no matter how alike the trauma, pain remains so hyper-personalised that it becomes fundamentally 'unsharable', 'incomprehensible' and 'unthinkable'. Scarry also argues that the world-altering nature of war exacerbates the 'un-making' character of trauma.[4] Arthur Kleinman goes further, drawing a clear line between 'real' (physical) pain and 'functional' (psychological) pain in victims of mass trauma.[5] From these perspectives, pain belongs only to those who experience it directly; observers and bystanders cannot possibly understand trauma and should not even try.

Yet, scholars regularly posit 'sharing' experiences as a prescription for the 'problem' of traumatic memory, and doing so widens the circle of pain. The true, lasting impact of trauma comes after (often long after) the actual incident. Unless dealt with, the pain of memory becomes 'ceaseless' and 'perennial'; an 'interminable death'. The guilt and shame of surviving alone can often *feel* more painful than the initial event. The suggestion is that by 'speaking out' and 'bearing witness' in art, in court and in writing, survivors can find cathartic release.[6] Victim communities likewise stress the importance of 'passing on' traumatic memory to future generations. Primo Levi, a Holocaust survivor and Nobel laureate, considered sharing pain the 'duty of the living' and 'imperative of awakening'.[7] Echoing Levi's call, Chinese author Ba Jin wrote of 'finding solace and release in pen and paper' in the wake of the devastating Cultural Revolution.[8] Consensus around bearing witness makes intuitive sense. However, the potential harm of doing so is less well understood and explored, both for the individuals affected and for those who receive and record the testimony. Historian Dominick LaCapra warns against dismissing 'secondary or muted trauma' too quickly, but the incidental distress of jurists and bystanders does not figure prominently in most discussions of pain and trauma.[9] With some justification, scholars worry that researching second-hand trauma implicitly equates it to the 'true', horrific suffering of victims and survivors. Universalising particular and personal feelings, and oversimplifying causal links between pain and responses is certainly a risk, but ignoring the actual experiences of international or humanitarian personnel on the ground overlooks a constitutive element of their work. Whether appropriate or not, I argue here that simply by investigating and hearing the cathartic release of survivors, participants in courts and related institutions become vulnerable to a new set of disruptive collateral emotions.

I am not the first to analyse the sharing of pain or how 'fragments of trauma' carry on through individuals and communities. An extensive literature probes the intergenerational transmission of Holocaust pain in families and classrooms.[10] Shoah scholarship has also delved into how its horrors – or the resulting guilt – transfer to perpetrator societies.[11] Increasingly, this gaze includes people recovering the 'lost' memory and pain of 'forgotten' mass atrocities. Shake Topalian, for example, describes the Armenian genocide's legacy as: 'Massive trauma transferred across the boundaries of generations ... How ghosts may be turned into ancestors.'[12] Similar developments explore the 'soul wound' caused by centuries of indigenous displacement in the Americas and Australia.[13] Family counsellors recognise that some combat veterans pass on war trauma to their children and spouses.[14] Research also demonstrates how trauma specialists and other people who work with pain frequently take on aspects of trauma from their hands-on experiences with the suffering of others. In the words of one study, 'traumatic experiences have a contagious nature'.[15] Indeed, since the groundbreaking work on the phenomenon of 'vicarious trauma' by Lisa McCann and

Laurie Anne Pearlman, psychologists have compiled a detailed – though overlapping – taxonomy of 'vicarious trauma', 'secondary trauma', 'compassion fatigue', 'burnout' and related conditions.[16] Visual representations of atrocity can be particularly damaging by searing pain onto public and personal consciousnesses. Works on the emotional valence of 'rubble photography' or 'atrocity photography' and other graphic 'conduits of trauma' reflect another way in which international criminal investigators become exposed.[17] Though important for building knowledge of how trauma 'shares' and for spreading information about treating such pain, few works have looked into how collateral emotions shape individuals and institutions of international justice and humanitarianism.

The emotional risks of justice work are under-researched, but not completely ignored. Karen Brounéus, for example, recently identified trauma transferral in Rwanda's *gacaca* courts. These restorative community-level 'courts' bring survivors, bystanders and perpetrators of Rwanda's 1994 genocide together in intimate settings.[18] At the trial's end, Amanda Gil, Matthew B. Johnson and Ingrid Johnson note the emotional difficulties met by correctional professionals and jurists involved in criminal executions.[19] Some Nuremberg and Tokyo participants, like the *gacaca* participants, struggled with the violent material with which they were confronted, and elsewhere I have demonstrated the strain Tokyo judges felt in exercising their life-or-death decision-making powers.[20] Certain conditions of international work intensified the risk and responses to IMT participation. First, intimate engagement with survivors and their memories heightened participant difficulties. Studies suggest that 'exposure to trauma survivors' terrifying, horrifying, and shocking images; strong, chaotic affect; and intrusive traumatic memories' harms individuals and 'social networks',[21] and that trauma professionals experience stronger reactions when 'exposed to the same community trauma as their clients'.[22] In Nuremberg and Tokyo, IMT participants shared actual social circles with many of the witnesses who attended court. Moreover, they lived in sites of war atrocity, to some extent in the same community of trauma as the victims and perpetrators in court. Participants who had fought in the war revisited its horrors; newly arrived civilian members saw stylised notions of the conflict rent asunder. Life as 'victors' in ravaged cities created sticky ethical issues. The Allies won, but at what cost to lives and values? Surrounded by the suffering and starkness of post-war recovery, participants contemplated the war's totality and brutality in court and out. Yet, at the same time, personnel at both IMTs actively shared in an ebullient period of ideals and internationalism imbued with an almost millenarian determination to ensure future world peace. The inevitable clash of principles and practices aggravated the emotional assault felt by some participants. An exaggerated sense of purpose combined with a culture of law built on the suppression of emotions exposed participants. By endeavouring to keep emotions 'out' of court and preserving unattainable justice ideals, the IMTs opened the door to let pain 'in' to participants' hearts and minds.[23]

Bearing witness; feeling pain

An epochal 'never again' mentality placed the IMTs and their participants alongside other related institutions in a formative international moment. The tribunals overlapped, *inter alia*: the first session of the UN General Assembly (January 1946), the initial hearings of the International Court of Justice (April 1946), the adoption of a Genocide Convention (December 1948) and the signing of the Universal Declaration of Human Rights (December 1948). Ultimately, few of these institutions lived up to expectations.[24] Long-term limitations, however, cannot erase the idealism and internationalism theorised, implemented and *felt* during the period. For many individual contributors to the movement, in Tokyo and elsewhere, the post-war era represented a time of change and optimism. Developments in Nuremberg, Tokyo and beyond looked like signal achievements in human history. A determination to change future outcomes and redress past wrongs emerged as an almost universal guiding principle at the IMTs. The omnibus construct of 'never again' itself grew from a wider post-war *Weltanschauung* expecting a revolution in global and human security. For many IMT participants, the 'new' world order lay in the creation of improved international and legal systems. 'In a sense with all the pressing problems which occupy us the trials [Nuremberg and Tokyo] may seem somewhat unnecessary, something which relates to the past', Lord Robert Alderson Wright (Chairman of the United Nations War Crimes Commission) told a special meeting of the Far Eastern Commission (FEC) in June 1946:

> But they are really related to the future ... These trials, no matter how imperfect they are, should be supported and justified. Really it is the future we are thinking about, not the past, for the past is beyond reparation. The only thing is not to let it happen again.[25]

There was no choice. As Nuremberg observer Rebecca West opined, 'a gaping hole would have appeared in our moral system had it been possible for villains to commit a vast number of vile crimes in their own and other countries, and to escape punishment'.[26]

Elevated expectations intensified ground-level responses, which made the disjuncture more jarring between lofty ideals and grim atrocities in court and in investigation rooms. The trials in action never embodied what participants had imagined coming in; the justice meted out was incomplete at best. 'The Tribunal, like the Apocalypse, was supposed to drive out evil and enthrone good', wrote Nuremberg prosecutor Telford Taylor. 'But the goal was not attained on four horses. For nearly a year the inmates and workers of the courthouse had been fairly drowned in documents, arguments, speeches, witnesses, translators, reporters, and other judicial whatnot.'[27] When romance met reality, the IMTs disappointed. Expecting justice on high, participants instead confronted 'Citadel[s] of boredom'.[28] Internal disillusionment went beyond doldrums. Participants felt *hurt* by the inability of the courts to live up to grand expectations. 'The Nuremberg trial must be admitted as a betrayal of the hopes that it engendered', wrote West. 'It was an unshapely event, a defective composition, stamping no clear image on the mind of the people it had been designed to impress.'[29] For some participants, the addition of unsettling images and experiences of violence to this inner welter turned mundane disappointment into emotional disquiet, even trauma.

Of course, alternation and trauma hardly formed a universal response to working in Nuremberg and Tokyo. Some employees considered the trials to be work: an assignment or a job to perform. Associated boredom was expected, even welcome, after a long and brutal war. For others, the employment became an adventure, an exciting life abroad. Still others approached their contribution as professional benchmarks; prominent conclusions to established careers or illustrious starts of future success. But some Nuremberg and Tokyo employees' experiences were unquestionably ones of opportunity and alternation. Ultimately, atrocities not only helped build the IMTs, but their place in court, in turn, transformed participants themselves. This 'personnel legacy' may indeed be one of the most important contributions of the IMTs. Unsurprisingly, perhaps, psychology literature on secondary trauma, burnout, compassion fatigue, vicarious trauma and related diagnoses tends to explore psychological responses by mental health caregivers, therapists and clinicians. Lessons from Nuremberg and Tokyo suggest the need to apply a more inclusive definition of pain and trauma to a wider array of professional spheres dealing in human suffering and feelings. Philip Kapleau, Harold Evans and others reveal how in certain situations the 'pain' felt in observing and confronting humanitarian crises engenders emotional, spiritual, moral, professional and even physical alternations that include but are by no means limited to psychological change. All this counts – or should count – as pain, even trauma, in the right conditions.

Words like 'trauma' and 'pain' hold purely negative connotations. Yet, pain can also be associated with growth; traumatic settings can open the door for agency and even opportunity. 'There is a cost to caring', writes trauma psychologist Noreen Tehrani, but evidence also exists of 'good emerging from tragedy', with certain people emerging from secondary trauma with 'an enriched philosophy on life'.[30] To experience traumatic growth, individuals must first feel pain strong enough to 'shatter ... existing beliefs' and provoke 'feelings of hopelessness, helplessness, impotence, and fear'.[31] After being broken down by 'grief' and 'numb[ness]', some individuals build new ways of thinking, feeling and acting.[32] Unlike trauma, which is normally understood to happen suddenly, traumatic growth can take a long time to emerge. When first confronted with violent images and testimonies, individuals may be unaware of the transformation taking place within. They may even feel satisfied and fulfilled with the job at hand, only to manifest change well after the fact. However, when alternation has occurred, the eventual manifestation becomes undeniable. Nuremberg and Tokyo court reporter Philip Kapleau and Tokyo judicial assistant Harold Evans both experienced classic cases of secondary trauma and post-traumatic growth from their time with the IMTs.

Kapleau and Evans

No one embodies the powerful link between observing pain, pain in observing and personal transformation quite like Philip Kapleau. Born in 1912, Kapleau attended Protestant, Catholic and Jewish services from a young age. He looked at faith with a mix of curiosity and scepticism. In his own words, he became 'first a freethinker, then an agnostic ... then an atheist'.[33] By the end of the Second World War, he had established himself as a well-regarded court reporter in Connecticut with little time or inclination for esoteric ruminations on metaphysics. He felt excited when the opportunity to work in Nuremberg presented itself. After completing the application process, including passing a series of physical and mental tests, he accepted a position as court reporter to the IMT in Nuremberg. Two things struck him upon his arrival in Germany: the devastation of the city and what he perceived to be the lack of repentance of the population. His disquiet worsened as he began work in court:

> The testimony at the trials was a litany of Nazi betrayal and aggression, a chronicle of unbelievable cruelty and human degradation. Listening day after day to victims of the Nazis describe the atrocities they themselves had been subjected to or had witnessed, one was shocked into numbness, the mind unable to comprehend the enormity of the crimes.[34]

Kapleau's emotional 'numbness' – a hallmark of secondary trauma and growth – became the beginning of a life-transformation. The experience plunged him into 'the deepest gloom' and the spiritual curiosity of his youth 'suddenly burst again into full consciousness'.[35]

Kapleau's responses highlight the immediacy of collateral emotions, the drawn-out nature of resulting pain, and the extent of social and personal disruption forced by secondary trauma. Though by no means on the same level of suffering experienced by victims of the atrocities he recorded, Kapleau's pain felt real and resulted in actual, enormous change. The court reporter's pain came on strong, within weeks of working in Germany. His transformation, however, played out over decades. Deeply unhappy and unsettled in Germany, he applied for a transfer to the newly formed Tokyo IMT. At least partly, he went to Tokyo for adventure. He had never travelled in Asia and looked forward to experiencing new cultures. Mostly, however, he left Nuremberg because he had reached 'a mood of black depression tinged with shame and guilt'.[36] 'At a deep level', he later explained, 'I felt somehow that I, too, was responsible for the overwhelming sufferings reverberating from the war. Japan could not be worse than Germany.'[37] On one level, Kapleau was wrong. In Tokyo he witnessed misery and suffering that was indeed 'worse' than Germany. What became apparent to him, however, was a striking contrast between Japan's acceptance of its post-war condition and the 'self-justifications' he heard in Germany.[38] When he asked acquaintances about this difference, he was told that the root of acceptance lay in the notion of 'karmic retribution'. Intrigued, he began asking questions about karma. His answer – figuratively and literally – came in the form of Dr D.T. Suzuki.[39]

Suzuki was a renowned Zen scholar, fluent in English and passionate about introducing Zen to the West. In a handful of meetings, Suzuki introduced Kapleau to ideas that would become the foundation of a philosophical and spiritual rebirth. On his return to Connecticut in

1948, Kapleau found that he could not satisfactorily settle back down into the routines of his former life. When Suzuki took up a position at Columbia University in 1950, Kapleau leapt at the opportunity to study under him. As Kapleau's engagement with Zen deepened, so too did his dissatisfaction with the 'middle class comforts and values' of his home life.[40] In 1953, the awakening that began with his experiences in Nuremberg and Tokyo reached a new level. He quit his job, left behind his old life and moved back to Japan, determined to dedicate himself to deeper levels of meditation and spiritual understanding.[41] For the next 12 years, he pursued intensive study under various Japanese Zen masters. After being ordained in 1965, 'Roshi' Kapleau began to yearn for home: 'I had grown stale and needed to renew myself through daily contact with the sights and sounds, the forms and customs of Western society.'[42] In 1965, he established the Rochester Zen Center in upstate New York. In the same year, he also published his first of many books. *The Three Pillars of Zen* remains a classic on the subject. It has been translated into 12 languages, including French, German, Italian and Spanish, and became one of the first English-language books to present Zen Buddhism not as philosophy, but as a pragmatic way of life.[43] Kapleau's works are noteworthy for their insights on war, peace and justice. *Awakening to Zen: The Teachings of Roshi Philip Kapleau* (1997), for example, reveals not just a Zen philosopher, but also a determined peace advocate. 'Massive war[s] of violence', he writes, 'have an adverse effect on all of us, provoking fear, anxiety, anger, depression, and other unhealthy emotions. It's obvious that in a physical and social sense, we need to work together to save our imperilled planet.'[44] Kapleau's life and work serve as powerful reminders of the destructive but also constructive nature of trauma and pain. His negative early confrontation of collateral emotions changed over time into a poignant example of post-traumatic growth.

Like Kapleau, Harold Evans witnessed pain all around him. Arriving in Tokyo in early 1946 as part of the New Zealand contingent to the Tokyo IMT, the young special assistant felt stunned by residual evidence of vicious Allied fire-bombings of Tokyo and its environs. Guilt complicated his feelings. He was not supposed to feel sympathy for the Japanese – he was supposed to feel victory. Yet the suffering that affected him, the pain he observed and the ache it caused came from the trauma of enemies. Beyond the moral ambiguities of modern war, Evans' response to Tokyo speaks volumes about the mutability and subjectivity of pain in history and life. Stunned by the aftermath of violence that surrounded him, Evans felt more troubled by 'how you get used to seeing [devastation], and how quickly you become adjusted to expecting to see it'. Daily exposure numbed him to 'whatever it was [I] did feel at first. There is no longer any conscious reaction to it'.[45] The New Zealander's correspondence files reveal the dystopic contrast between ideals and harsh realities among IMT participants in post-conflict zones:

> Housed in the extremely commodious Canadian Legation, and with office quarters in the undamaged Meiji Building in the central part of Tokyo, and being driven to and fro twice a day between these two places in cosy American sedans – [I remember] walking back to the Legation one snowy Sunday afternoon from the Hibiya Hall after the first of the Nippon Philharmonic concerts that I went to, and realising (I think) something of the tragedy of the destruction of Tokyo. From Hibiya Park to the Canadian Legation is a forty minute walk, and at walking pace you appreciate so much more than you can at 40 m.p.h. what this destruction has meant to the people who still live in the midst of it.[46]

Evans lived in expat luxury, while the Japanese around him choked on the war's ashes. 'Don't mistake these comments as being upon the so-much-debated questions of bombing of civilians or not or war or not', he reassured family and friends. 'I'm not on that ground.'[47] But he was, and he could not help admitting to feeling 'sadness', 'depression' and 'dumbfoundedness' at the scenes he witnessed.[48] Seeing first-hand the devastation of Hiroshima revealed an even greater 'tragedy'. The death and destruction seemed more tragic in conjunction with the artificial sterility of international judicial proceedings. Though moved by gruesome testimonies in court of Japanese crimes, the young New Zealander hated the double-standard of purportedly even-handed 'justice'. In a 2004 interview, he remembered feeling 'disappointed' by the obstinate silence of international justice in Tokyo regarding the nuclear attacks: 'The big queries came when the defence counsel tried to introduce the dropping of the bomb, you see. It was always ruled out of order by the court but it was not allowed to be mentioned, not even to be mentioned!' The ghosts of Hiroshima, Nagasaki, Tokyo and other bombed cities haunted Evans with their absence from the Tokyo IMT.

Is what Evans felt 'pain'? Can we equate a troubled conscience and galvanised spirit to 'trauma'? In the narrowest physical sense, perhaps Evans' feelings do not qualify. Yet, his transformative *response* to the agony around him holds distinctive hallmarks of trauma recovery. Research on traumatic growth suggests that people who feel they are doing good suffer less. People who doubt the value of their work become more likely to experience disruption. Although his 'sadness' spiked immediately, Evans' alternation emerged incrementally. Ultimately, his change manifested itself as positive growth, but it surfaced first as disillusionment. Early doubts about the IMT's fairness festered. High ideals dictated that true justice should be universal. The trauma seen in court mattered, but so too should the pain Evans observed all around him. All pain should be shared and salved. In Evans' eyes, the Tokyo IMT came to represent selective victors' justice; a betrayal of both personal sensibilities and traumatic realities perceived on the ground.[49] Determined to stand for the best hopes espoused in Tokyo, not for its shortcomings, Evans became an active humanitarian. In the 1960s and 1970s, he contributed to New Zealand's activist community against nuclear testing in the Pacific. In the 1980s and 1990s, he played a central role in a transnational advocacy network that eventually forced the International Court of Justice officially to condemn the threat and use of nuclear weapons.[50] Rooted in his life-altering confrontation of Japanese pain, his later activism reminds us of the human dimension to secondary trauma and traumatic growth. Together with Philip Kapleau, Evans represents the powerful emotions and personnel legacies of the Nuremberg and Tokyo IMTs. More broadly, his experiences highlight the complex forces and forms of pain.

Conclusions

Emotional responses created living tribunal legacies. The experiences of some IMT employees followed a pattern of opportunity and alternation, trauma and growth. Employment at the IMTs provided an exciting personal and professional *opportunity* for those involved, particularly for young participants. Administering international justice in post-conflict societies, however, comes with a number of inherent difficulties. In court and out, IMT employees in Nuremberg and Tokyo regularly faced graphic and disturbing evidence of Axis and Allied atrocities. Disjunction from the comforts and mores of 'home' unsettled many participants, though many also felt swept up in the exuberant ideals at work. For some, the push-pull of unbounded idealism and emotional agitation led to an *alternation* experience. In short, working for the IMTs became a life-changing event. The stark experience of war and the heady high stakes of post-conflict environments crystallised personal ideals and character. This facet manifested itself at superficial levels like language, but also provoked profound psychological and emotional shifts. The messy realities – sometimes surrealities – of war and its aftermath force and engender psychological and intuitive alterations, especially in total wars couched heavily in competing visions of purity and exceptionalism. The brutality, bravado and righteousness of the Second World War sharpened, and in some cases shattered, personal worldviews. For individuals like Philip Kapleau and Harold Evans, observing pain formed an unhinging experience, but it ultimately manifested itself as one of growth and transformation.

The individual emotional experiences of Kapleau and Evans suggest broader conclusions about the personal difficulties and outcomes of other global humanitarian, justice and governance projects. Where this chapter focuses on collateral emotions in international courts, it identifies a much broader trend in humanitarian and post-conflict arenas – to say nothing of everyday emergency response, trauma therapy, sexual violence counselling, social work and other high-risk spaces more generally. Though not always identified by other researchers in terms of secondary trauma, collateral emotions or traumatic growth, examples abound of historical actors finding pain in the suffering of others – and undergoing powerful personal transformations as a result. Human rights scholar Samantha Power captures the anguish and agony of genocide bystanders in her seminal *A Problem from Hell: America in the Age of Genocide*. She presents particularly evocative accounts of Henry Morgenthau's time as the US Ambassador to the Ottoman Empire during the Armenian genocide and of Canadian General Roméo Dallaire's failed interventions during the Rwandan genocide. John Hagan describes the 'alternation' experience of Nuremberg prosecutor Benjamin Ferencz, who spent the rest of his life as an international legal advocate.[51] Likewise, Peter Balakian's writings on his great-grandfather Grigoris Palak'ean's response to the Armenian genocide powerfully illustrate the transformation and pain of witnessing trauma.[52] Biographical studies of Raphael Lemkin's early genocide advocacy show much the same.[53] Individuals experienced and observed pain in unique ways. Though never equal to the trauma of victims and survivors, the individual responses explored here speak to real pain felt by actors in the high-stakes world of international justice, humanitarianism and atrocity. Their lives tell us that feelings can operate as historical actors in and of themselves.

Notes

1. I would like to thank colleagues in the Department of History & Classics at Acadia University, the Strassler Centre for Holocaust & Genocide Studies at Clark University and the Department of History at the University of

British Columbia for general help in building this chapter and its ideas. Conversations with Nick Bogod, clinical faculty member in Neurology at the Department of Medicine at the University of British Columbia, introduced me to a world of trauma and growth research. Finally, I am grateful to Rob Boddice, the Birkbeck Pain Project and the Wellcome Trust for bringing together such an interesting group of experts on pain and emotions.

2. Elaine Scarry, *The Body in Pain: The Making and Unmaking of the World* (New York: Oxford University Press, 1985), 112; and Feng Jicai, *Ten Years of Madness: Oral Histories of China's Cultural Revolution* (trans. Dietrich Tschanz) (San Francisco: China Books & Periodicals, Inc., 1996), 273.
3. See Cathy Caruth, *Unclaimed Experience: Trauma, Narrative, and History* (Baltimore: Johns Hopkins University Press, 1996); Lawrence L. Langer, *Preempting the Holocaust* (New Haven: Yale University Press, 1998); and Primo Levi, *The Drowned and the Saved* (trans. Raymond Rosenthal) (New York: Summit Books, 1988).
4. Scarry, *The Body in Pain*.
5. Arthur Kleinman, *Writing at the Margin: Discourse between Anthropology and Medicine* (Berkeley: University of California Press, 1995).
6. See note 3. See also, interweaving artistic and academic approaches to write the history of trauma, Shoshana Feldman, 'Education and Crisis, of the Vicissitudes of Teaching', in Cathy Caruth (ed.), *Trauma: Explorations in Memory* (Baltimore: Johns Hopkins University Press, 1995), 13–60; Dominick LaCapra, 'Holocaust Testimonies: Attending to the Victim's Voice', in Moishe Postone and Eric Santner (eds), *Catastrophe and Meaning: The Holocaust and the Twentieth Century* (University of Chicago Press, 2003), 209–31; Vera Schwarcz, 'The "Black Milk" of Historical Consciousness: Thinking About the Nanking Massacre in Light of Jewish Memory', in Fei Fei Li, Robert Sabella and David Liu (eds), *Nanking 1937: Memory and Healing* (Armonk, NY: M.E. Sharpe, 2002), 183–205; Rubie Watson, 'Memory, History, and Opposition under State Socialism: An Introduction', in Rubie S. Watson (ed.), *Memory, History, and Opposition under State Socialism* (Santa Fe, NM: School of American Research Press, 1994), 1–20; Froma Zeitlin, 'New Soundings in Holocaust Literature: A Surplus of Memory', in Postone and Santner (eds), *Catastrophe and Meaning*, 173–208.
7. Levi, *The Drowned and the Saved*.
8. Ba Jin, *Random Thoughts* (trans. Geremie Barmé) (Hong Kong: Joint Publishing Company, 1984), 21.
9. LaCapra, 'Holocaust Testimonies', 220.
10. Families: Mildred Antonelli, 'Intergenerational Impact of the Trauma of a Pogrom', *Journal of Loss and Trauma*, 17(4) (2012): 388–401; Nanette C. Auerhahn and Dori Laub, 'Intergenerational Memory of the Holocaust', in Yael Danieli (ed.), *International Handbook of Multigenerational Legacies of Trauma* (New York: Plenum Press, 1998), 21–41; Yael Danieli, 'The Treatment and Prevention of Long-Term Effects and Intergenerational Transmission of Victimization: A Lesson from Holocaust Survivors and their Children', in Charles R. Figley (ed.), *Trauma and its Wake* (Levittown, PA: Brunner/ Mazel Publishers, Inc., 1985), 295–313; John J. Sigal and Morton Weinfeld, *Trauma and Rebirth: Intergenerational Effects of the Holocaust* (New York: Praeger, 1989); Hadas Wiseman, Einat Metzl and Jacques P. Barber, 'Anger,

Guilt, and Intergenerational Communication of Trauma in the Interpersonal
Narratives of Second Generation Holocaust Survivors', *American Journal of
Orthopsychiatry*, 76(2) (2006): 176–84. Classrooms: Katherine Bischoping,
'*Timor Mortis Conturbat Me*: Genocide Pedagogy and Vicarious Trauma', *Journal
of Genocide Research*, 6(4) (2004): 545–66; and Felman, 'Education and Crisis'.

11. Jost Dülffer, 'The Adenauer Era – Anxieties and Traumas of Violence in the
Postwar Period', *German Politics and Society*, 25(2) (2007): 1–6; Evelin Gerda
Lindner, 'Were Ordinary Germans Hitler's "Willing Executioners"? Or Were
they Victims of Humiliating Seduction and Abandonment? The Case of
Germany and Somalia', *IDEA: A Journal of Social Issues*, 5(1) (2000); and
Evelin Gerda Lindner, 'Humiliation-Trauma that Has Been Overlooked: An
Analysis Based on Fieldwork in Germany, Rwanda/Burundi, and Somalia',
Traumatology, 7(1) (2001): 43–68. Tsutsui Kiyoteru has similarly explored the
'collective trauma of perpetration' in Japanese war guilt. Tsutsui Kiyoteru,
'The Trajectory of Perpetrators' Trauma: Mnemonic Politics Around the
Asia-Pacific War in Japan', *Social Forces*, 87(3) (2009): 1389–422.

12. Shake Topalian, 'Ghosts to Ancestors: Bearing Witness to "My" Experience
of Genocide', *International Journal of Psychoanalytic Self Psychology*, 8(1)
(2013): 7–19.

13. Maria Yellow Horse Brave Heart and Lemyra M. DeBruyn, 'The American
Indian Holocaust: Healing Historical Unresolved Grief', *American Indian
and Alaska Native Mental Health Research: Journal of the National Center*, 8(2)
(1998): 56–78; Eduardo Duran, Bonnie Duran, Maria Yellow Horse Brave
Heart and Susan Yellow Horse-Davis, 'Healing the American Indian Soul
Wound', in Danieli (ed.), *International Handbook*, 341–54; and Beverley
Raphael, Patricia Swan and Nada Martinek, 'Intergenerational Aspects of
Trauma for Australian Aboriginal People', in Danieli (ed.), *International
Handbook*, 327–39.

14. See, for example, Joseph R. Herzog, R. Blaine Everson and James D. Whitworth,
'Do Secondary Trauma Symptoms in Spouses of Combat-Exposed National
Guard Soldiers Mediate Impacts of Soldiers' Trauma Exposure on their
Children?', *Child & Adolescent Social Work Journal*, 28(6) (2011): 459–73;
Robert W. Motta, Jamie M. Joseph, Raphael D. Rose, John M. Suozzi
and Laura J. Leiderman, 'Secondary Trauma: Assessing Inter-generational
Transmission of War Experiences with a Modified Stroop Procedure', *Journal
of Clinical Psychology*, 53(8) (1997): 895–903; Michael Weinberg, 'Spousal
Perception of Primary Terror Victims' Coping Strategies and Secondary
Trauma', *Journal of Loss and Trauma*, 16(6) (2011): 529–41.

15. Robert W. Motta, 'Secondary Trauma in Children and School Personnel',
Journal of Applied School Psychology, 28(3) (2012): 256–69 at 266.

16. I. Lisa McCann and Laurie Anne Pearlman, 'Vicarious Traumatization: A
Framework for Understanding the Psychological Effects of Working with
Victims', *Journal of Traumatic Stress*, 3(1) (1990): 131–49. See also Priscilla
Dass-Brailsford and Rebecca Thomley, 'An Investigation of Secondary Trauma
Among Mental Health Volunteers after Hurricane Katrina', *Journal of Systemic
Therapies*, 31(3) (2012): 36–52; Grant J. Devilly, Renee Wright and Tracey
Varker, 'Vicarious Trauma, Secondary Traumatic Stress or Simply Burnout?
Effect of Trauma Therapy on Mental Health Professionals', *Australian and
New Zealand Journal of Psychiatry*, 43(4) (2009): 373–85; Sharon Rae Jenkins

and Stephanie Baird, 'Secondary Traumatic Stress and Vicarious Trauma: A Validational Study', *Journal of Traumatic Stress*, 15(5) (2002): 423–32; Brian E. Perron and Barbara S. Hiltz, 'Burnout and Secondary Trauma Among Forensic Interviewers of Abused Children', *Child and Adolescent Social Work Journal*, 23(2) (2006): 216–34; Mary Dale Salston and Charles R. Figley, 'Secondary Traumatic Stress Effects of Working with Survivors of Criminal Victimization', *Journal of Traumatic Stress*, 16(2) (2003): 167–74. Some scholars criticise the 'unsubstantiated' claims about secondary trauma. Michaela Kadambi and Liam Ennis note that: 'The majority of professionals are *not* suffering emotional or psychological distress in response to clinical work and are coping well with the demands of their work.' Michaela A. Kadambi and Liam Ennis, 'Reconsidering Vicarious Trauma', *Journal of Trauma Practice*, 3(2) (2004): 1–21. See also Lisa S. Elwood, Juliette Mott, Jeffrey M. Lohr and Tara E. Galovski, 'Secondary Trauma Symptoms in Clinicians: A Critical Review of the Construct, Specificity, and Implications for Trauma-Focused Treatment', *Clinical Psychology Review*, 31(1) (2011): 25–36. Majority or not, the responses in this chapter represent an important and overlooked dimension of international justice.

17. Steven Hoelscher, '"Dresden, a Camera Accuses": Rubble Photography and the Politics of Memory in a Divided Germany', *History of Photography*, 36(3) (2012): 288–305; Christina Twomey, 'Framing Atrocity: Photography and Humanitarianism', *History of Photography*, 36(3) (2012): 255–64; Rick Crownshaw, 'Photography and Memory in Holocaust Museums', *Mortality*, 12(2) (2007): 176–92. See also Marianne Hirsch, *Family Frames: Photography, Narrative, and Postmemory* (Cambridge, MA: Harvard University Press, 1997); Marianne Hirsch, 'Surviving Images: Holocaust Photographs and the Work of Postmemory', *Yale Journal of Criticism*, 14(1) (2001): 5–37.

18. Karen Brounéus, 'The Trauma of Truth Telling: Effects of Witnessing in the Rwandan *Gacaca* Courts on Psychological Health', *Journal of Conflict Resolution*, 54(3) (2010): 408–437; Karen Brounéus, 'Truth-Telling as Talking Cure? Insecurity and Retraumatization in the Rwandan *Gacaca* Courts', *Security Dialogue*, 39(1) (2008): 55–76.

19. Amanda Gil, Matthew B. Johnson and Ingrid Johnson, 'Secondary Trauma Associated with State Executions: Testimony Regarding Execution Procedures', *Journal of Psychiatry & Law*, 34 (2006): 25–35.

20. The Dutch judge in Tokyo wrote: 'Life is pretty difficult and unpleasant ... I sit at my desk wondering whether someone has to be hanged or to be shot, which is in the long run rather depressive.' B.V.A. Röling to Delphine (25 June 1948), Archief van B.V.A. Röling - 2.21.273, *Nationaal Archief Den Hague*, *Bestanddeel* 27 (hereinafter 'Röling Papers'), Box 27. In private conversation, Supreme Commander for the Allied Powers (SCAP) General Douglas MacArthur confessed that signing the Tokyo IMT's death sentences was 'the hardest job of his life'. Sir Patrick Shaw, Allied Council for Japan to Department of External Affairs, Wellington (15 November 1948), Archives New Zealand, Wellington, New Zealand. EA2 1948-29A 106-3-22 Part 8 (hereinafter 'NZ Archives'). See also James Burnham Sedgwick, 'A People's Court: Emotion, Participant Experiences, and the Shaping of Postwar Justice at the International Military Tribunal for the Far East, 1946–1948', *Diplomacy & Statecraft*, 22(3) (2011): 480–99. Rebecca West marked a similar air in

Nuremberg: 'Though it might be right to hang these men, it was not easy. A sadness fell on the lawyers engaged in the trial', she wrote shortly after the trial. 'Now this day of judgment had come they were not happy. There was a gloom about the places they lived, a gloom about their families ... If a trial for murder lasts too long, more than the murder will out. The man in the murderer will out; it becomes horrible to think of destroying him.' Rebecca West, *A Train of Powder* (New York: Viking Press, 1955), 41–3.

21. Jenkins and Baird, 'Secondary Traumatic Stress', 423.

22. Carol Tosone, Orit Nuttman-Shwartz and Tricia Stephen, 'Shared Trauma: When the Professional is Personal', *Clinical Social Work Journal*, 40(2) (2012): 231.

23. For the stunted emotions of legal proceedings, see Terry A. Maroney, 'Law and Emotion: A Proposed Taxonomy of an Emerging Field', *Law and Human Behavior*, 30(2) (2006): 119–42; James Burnham Sedgwick, 'Memory on Trial: Constructing and Contesting the "Rape of Nanking" at the International Military Tribunal for the Far East, 1946–1948', *Modern Asian Studies*, 43(5) (2009): 1229–54; Richard L. Wiener, Brian H. Bornstein and Amy Voss, 'Emotion and the Law: A Framework for Inquiry', *Law and Human Behavior*, 30(2) (2006): 231–48; Olga Tsoudis and Lynn Smith-Lovin, 'How Bad was it? The Effects of Victim and Perpetrator Emotion on Responses to Criminal Court Vignettes', *Social Forces*, 77(2) (1998): 695–722.

24. Mark Mazower and others question both the effectiveness and novelty of the era's international institutionalisation. Mark Mazower, *No Enchanted Palace: The End of Empire and the Ideological Origins of the United Nations* (Princeton University Press, 2009).

25. Extract from H.M.G. Document, Minutes of Special Meeting of FEC (16 June 1946), in NZ Archives, EA2 1946-31B 106-3-22 Part 1.

26. West, *A Train of Powder*, 245.

27. Telford Taylor, *Anatomy of the Nuremberg Trials: A Personal Memoir* (New York: Knopf, 1992), 546.

28. West, *A Train of Powder*, 3.

29. *Ibid.*, 246.

30. Noreen Tehrani, 'The Cost of Caring: The Impact of Secondary Trauma on Assumptions, Values and Beliefs', *Counselling Psychology Quarterly*, 20(4) (2007): 325 and 328.

31. Allysa J. Barrington and Jane Shakespeare-Finch, 'Working with Refugee Survivors of Torture and Trauma: An Opportunity for Vicarious Post-Traumatic Growth', *Counselling Psychology Quarterly*, 26(1) (2013): 90.

32. Tehrani, 'The Cost of Caring', 328.

33. Roshi Philip Kapleau, *Zen: Dawn in the West* (Garden City, NY: Anchor Press/Doubleday, 1979), 260.

34. *Ibid.*, 261.

35. *Ibid.*

36. *Ibid.*

37. *Ibid.*

38. In his seminal work of the same title about the Allied occupation of Japan, John Dower calls Japan's acceptance of the post-war world 'embracing defeat'. John W. Dower, *Embracing Defeat: Japan in the Wake of World War II* (New York: W.W. Norton & Co., 1999).

39. Kapleau, *Zen*, 262.
40. *Ibid.*, 259.
41. *Ibid.*, 264–5.
42. *Ibid.*, 268.
43. Philip Kapleau, *The Three Pillars of Zen: Teaching, Practice, and Enlightenment* (New York: Harper & Row, 1966).
44. Roshi Philip Kapleau, with Polly Young-Eisendrath and Rafe Martin (eds), *Awakening to Zen: The Teachings of Roshi Philip Kapleau* (New York: Scribner, 1997), 228.
45. Harold Evans to Family and Friends (15 June 1946), Harold Evans Papers, Macmillan Brown Library Archives Collection, University of Canterbury, Christchurch, New Zealand. MB 1559, Box 16, Folder 1. Hereinafter, 'Evans Papers'.
46. *Ibid.*
47. *Ibid.*
48. *Ibid.*
49. The selective justice he witnessed in Tokyo informed the New Zealander's later advocacy. 'The positive, absolute need is for a single all-restraining, universal outlawing of nuclear weapons, universally applied', he later wrote. 'The World Court Project aims to affirm life. It attempts to outlaw nuclear weapons of mass destruction and extermination of life, and in the process to deal a death blow to war itself.' 'Some Thoughts on the World Court Project' (29 April 1994), Evans Papers, Box 7, Folder 1. See also Harold Evans, 'The World Court Project on Nuclear Weapons and International Law', *New Zealand Law Journal* (1993): 249–52.
50. Kate Dewes, 'Taking Nuclear Weapons to Court', in Fredrik S. Heffermehl (ed.), *Peace is Possible* (Geneva: International Peace Bureau, 2000). See also Kate Dewes and Robert Green, 'The World Court Project: How a Citizen Network Can Influence the United Nations', *Social Alternatives*, 15(3) (1996): 35–7; Kate Dewes, *The World Court Project: The Evolution and Impact of an Effective Citizens' Movement* (Christchurch, NZ: Disarmament & Security Centre, 1998).
51. John Hagan, *Justice in the Balkans: Prosecuting War Crimes in the Hague Tribunal* (University of Chicago Press, 2003).
52. Peter Balakian, *Black Dog of Fate: A Memoir* (New York: Basic Books, 1997); Peter Balakian, *The Burning Tigris: The Armenian Genocide and America's Response* (New York: HarperCollins, 2003); Grigoris Palak'ean, *Armenian Golgotha* (trans. Peter Balakian and Aris G. Sevag) (New York: Alfred A. Knopf, 2009).
53. John Cooper, *Raphael Lemkin and the Struggle for the Genocide Convention* (New York: Palgrave Macmillan, 2008); Dan Eshet, *Totally Unofficial: Raphael Lemkin and the Genocide Convention* (Brookline, MA: Facing History and Ourselves, 2007); James Joseph Martin, *The Man Who Invented 'Genocide': The Public Career and Consequences of Raphael Lemkin* (Torrance, CA: Institute for Historical Review, 1984).

15
Documenting Bodies: Pain Surfaces

Johanna Willenfelt

This chapter sets out to explore the representability of physical pain in inter-embodied relationships and encounters. It asks how the emotive, transient and transformative space of contemporary art might contribute to new understandings of the notional experience of *sharing pain*.

In this text, I want to ask what would happen if we were to consider pain neither as an effect nor an affect, but rather tried to grasp it as an emotive, almost tangible object in a world replete with other objects. To this end, I will be looking into new ways of thinking about the agency of bodies and objects – animate as well as inanimate – and the implications of the notional non-representational technique of *allure* for acts of observing and interpreting the pain of others. I will be drawing from heuristic and pragmatic approaches from the field of visual art as well as looking into ideas from a new current in continental philosophy, referred to as the *speculative turn*, including an object-oriented theory of relations.

Primarily, I will be thinking through and with my own artistic practice and the recent project *Documenting Bodies* (2010), which explored the sociality and representability of pain through the activation of medical and surgical archives (Figure 15.1). *Documenting Bodies* was a research-based work that emerged through the interplay between the medical sphere and the artist's studio. Taking an interest in the journaling of medical doctors, the project served as a platform for interdisciplinary investigation, merging first-person accounts and methods with pain studies in an expanded field.[1] Throughout, the practice-led research process was largely informed by the context of medical history and doctor-patient relationships from the first half of the twentieth century.

Although it is important to distinguish pain from its interrelated area of suffering, not least on account of ethics, for the purpose of this study,

260

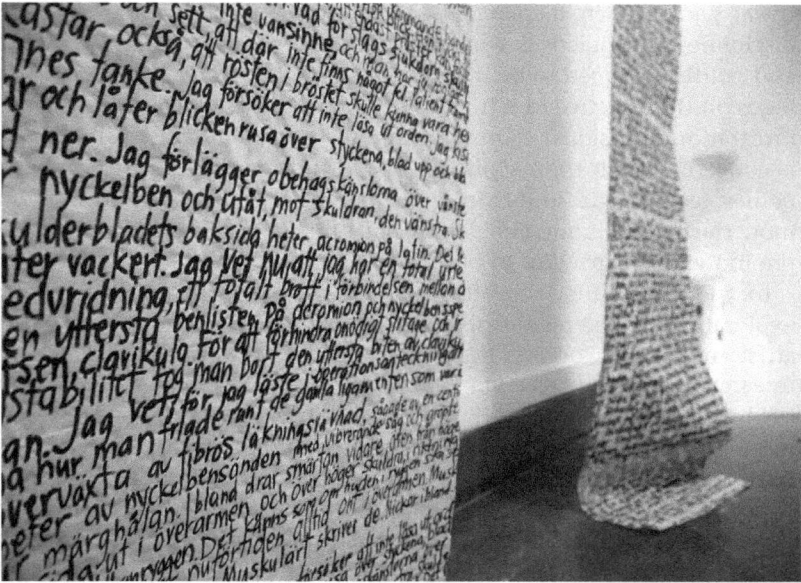

Figure 15.1 J. Willenfelt, *Documenting Bodies* (1) (2010)

pain and suffering form two separate objects with infrangible realities. Concurrently, as concepts, they are intertwined. Therefore, in this text, the notions of pain and suffering will be used alternately. Moreover, medical findings show that pain centres and emotion-processing regions neurally overlap, making it impossible to differentiate between physical and emotional components of pain on a practical level.[2] The pain-emotion contingency corroborates the obsoleteness of a mind-body dualism. Accordingly, we can have feelings of pain without bodies (phantom pain) and bodies in pain without feelings of physical pain, which is the case in congenital analgesia.[3] Yet, both conditions are unquestionably associated with suffering.

Affecting spaces

Spaces of contemporary art are inherently places of spatial, grammatical and affective transformation. Historically, emotionally charged spaces have been looked upon suspiciously, from Plato's ban on poets to the *affective fallacy* in literary criticism in modern times. The *affective turn* in the arts and academia emerged as a result of a consolidation of different disciplines and practices where separate interests in the bodily, social and political seemed to be interlinked through affect. Significantly, the focus on the affective in scholarly work has not only dealt with the disintegration of residual Cartesianism, but has also theorised how affect is concerned with the reconfiguration of the interconnectivity among bodies, technologies and inanimate matter, and as a related notion how modern-day economic circulation has assisted in subjugating affect as pre-individual capacities to biopolitical control.[4]

In the visual arts, the affective turn is above all associated with explorations of the relationship among the bodies of the audience, the artist and the artwork. As the creative arts are now forming their own research field within academia (I am especially referring to the emergent field of artistic research in Scandinavia), attempts at theorising affect in new ways through practice-led research are underway. According to philosopher Gilles Deleuze, the artwork is concerned with the creation of *percepts* and affects, which together form a sensation in the perceiver. Percepts and affects are independent entities that belong to a reality outside the experience of the individual.[5] That is to say, affect as an agency or force belongs to the meeting and should not be equated with the affections that an individual undergoes.

Documenting Bodies: the making of the work

The modern experience of pain begins with the organisation of treatments alongside a commodification of medical institutions and technical facilities in the twentieth century. With the arrival of modern medicine, fewer people were left to languish with various illnesses in their own homes. On the flipside, the entry of specialised language alienated the patient, whose pain was now interpreted by medical expertise as a symptom.[6] Medical devices and diagnostic tests came to replace a large part of the patient-doctor interaction. Replacing a jointly constructed illness narrative, the truth of pain was now located in the doctor's story, against which all other stories were measured and judged 'true or false'.[7] The subsumption under the emergent techno-scientific regime deprived patients (and to a certain degree also the bodies of doctors) of their autonomy. Ultimately, the reification processes in modern medicine, as a part of the beginning of the late capitalist era, affected not only the ownership of a patient's illness story, but also the patient's sense of self.[8]

Through investigations into the medical and surgical archives of the General and Sahlgrenska Hospital in Gothenburg, I came across patient files that were no longer classified, dating from the 1930s. Narrated by medical doctors, surgeons and post-mortem examiners, the journals contained the imprints of other people's hurting bodies – suffering, aching, sore, feverish, sliced, drained, carried, broken bodies. My own experience of suffering from chronic pain, of vacillating between expectations, hope and despair, was in those moments in that archive there with the hurting realities of others, although I could not reach them. From numerous entries I discerned some 20 cases, predominantly concerning female patients (it should be noted that the *Documenting Bodies* project dealt first and foremost with femininely coded pain, that is to say, low-intensive, long-term pain and society's conception of it). In discerning affective encounters, I was focused on interpersonal and embodied gestures concerned with obscure feelings of pain, as opposed to pain referring to established signs of illness. Adapted, these gestures and practices were then relocated and, together with other sensual elements, fused into hybrid bodies. Shown

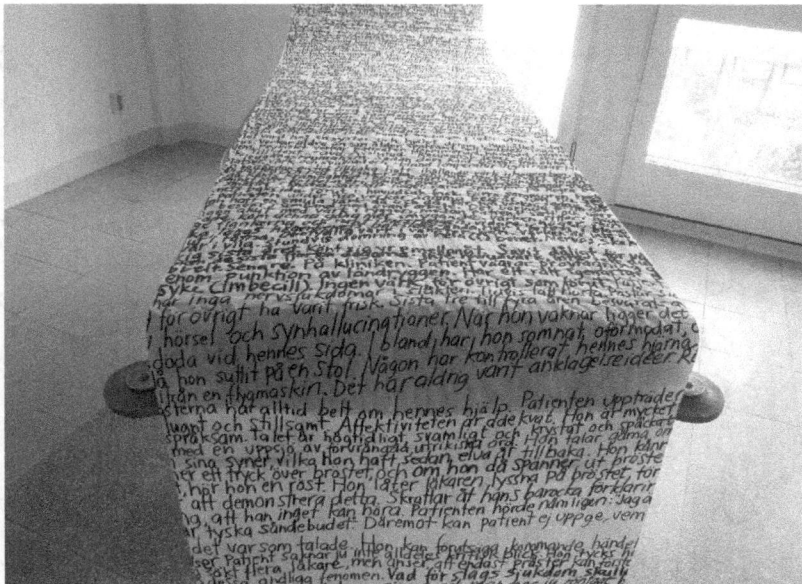

Figure 15.2   J. Willenfelt, *Documenting Bodies* (2) (2010)

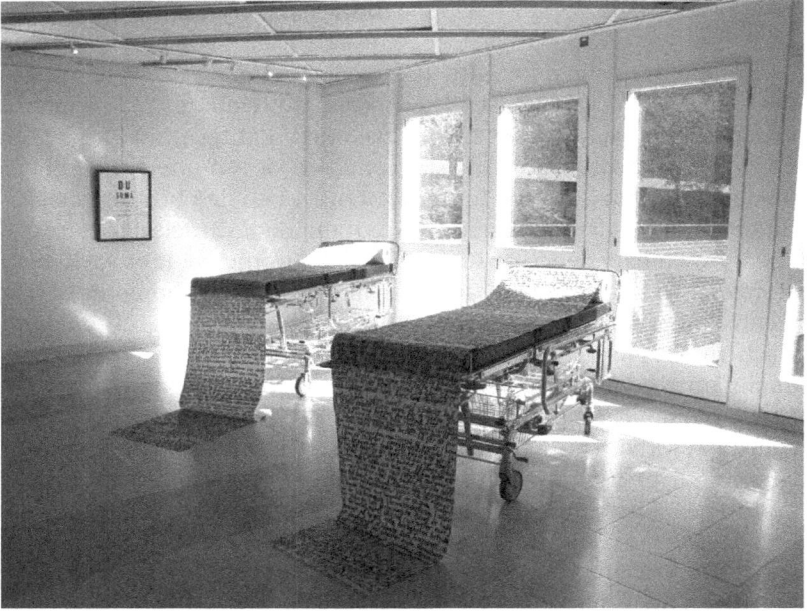

Figure 15.3 J. Willenfelt, *Documenting Bodies* (3) (2010)

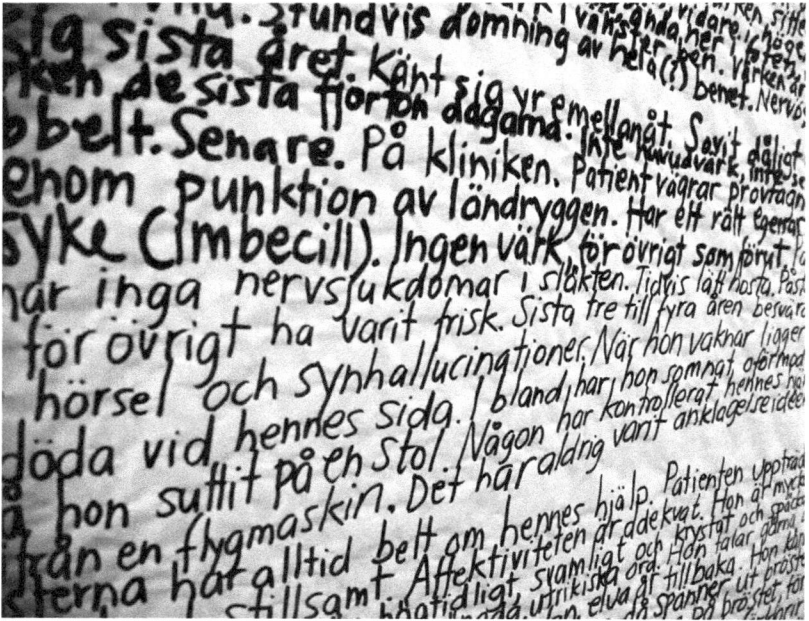

Figure 15.4 J. Willenfelt, *Documenting Bodies* (4) (2010)

in gallery contexts as well as in public spaces, hospitals and a county council, the hybrids contributed to more extensive installations comprised of drawings and texts as well as sonic objects and mixed media (Figures 15.2, 15.3 and 15.4).

The above image (Figure 15.4) is an excerpt of a hybrid text comprised of a number of interwoven medical opinions telling of statuses of patients, from physical appearances and individual features to emotional and affective responsiveness. As the text unfolded, so did a series of statements and testimonies. Different voices seemed to be making themselves heard: there was the patient, the nurse, the surgeon and what appeared to be a clinical assessment of sorts.

The method I used in *Documenting Bodies* I refer to as *re-contextualisation*. Re-contextualisation within the arts can be described as different approaches concerned with re-constitution of real, historical or contemporary events. The method denotes the way a specific and existing topic is garnered and activated by inclusion in a specific interrelationship within the sphere of art.[9] Marcel Duchamp's canonised work *Fountain* (1917) is often used to describe how conceptual art was established. Duchamp brought an object from outside the sphere of art into the museum. Conceptualising it as art, Duchamp claimed that this mode of conduct transformed the object, a manufactured urinal, into a work of art, leaving its former function behind.[10] When deploying re-contextualisation or re-enactment techniques in the making of artworks today, one would indeed affirm that something does happen to the object when it is transferred from one context to another. While still maintaining traits of its former self, the object would today probably assume the role as transformer in its own right, mutually affecting the space of art or the specific features of the site in which it was placed. In what will be discussed below, we will have reason to get back to this kind of transformation of reality.

Documenting Bodies was a work concerned with the redeployment of certain practices and gestures of pain, which, deprived of their circumstantial qualities, together with other sensual elements were fused into hybrid bodies. The fact that I made use of the pain and emotional suffering of others as artistic material necessitated an ethical stance. I needed to consider how an image of pain in the capacity of *found object* (*objet trouvé*) is ultimately a lived experience that belonged to some-*body*, intrinsically resistant to being replaced by representation. Notably, the perennial dilemma that emerges when bringing up the phenomenon of pain, in any context, is how to bridge the gap between self and other, human or otherwise. The perception of one's own embodiment separates one body from the next, making lived experiences of others hard to comprehend. There is always an imminent risk that one deals with a pain that one cannot feel according to the model of a pain one does feel or know as one's own, thus taking one's own experience as a starting point.[11] Even though I referenced my own body in video and text in the project, as an artist I inevitably take on the role of interpreter of the pain of others, with the impending risk of appropriation. In spite of the fact that medical narratives in the form of doctors' charts and diaries are great sources for understanding the prevailing scientific rationale and emotional conventions of a certain time, the retrieval of historical material in my project was not made as an attempt at restoring voices from the past. If voices from the past are to be 'heard' and redeemed, they must be so in a just way, in the right context, not to be capitalised upon or to become objects for one's own sadness, fear or hurt. In an attempt at recasting the collected gestures, I divested the statements and testimonies of their historical contexts by bringing them quite directly into a contemporary gaze (although I tried to keep untouched the specificity of each and every personal pain). Thus, the textual content in the work is a compound, which, apart from adaptations of filed medical journals, is also interspersed with literary fragments and the statements of others, as well as cut-outs from my own personal medical records. Consequently, the artwork is a highly morphed and speculative chronicle of a range of unidentified emotive objects of 'shared' pain encounters. What was acted upon in the work was not the conventions of a certain time or event in the past, but the contemporary context into which the objects of pain were redeployed as a composite event. Contemporary art can indeed unlock and draw attention to existing commodifications of pain and act for new ways to define oneself as a perceiving being in the world, but with it comes responsibility. As a *real object*, pain cannot be entirely seen or heard, only translated.

The visceral theory of emotion

If the affective turn legitimised feelings as study objects in postmodern, academic discourse, in Western philosophy William James is by many considered to be the thinker who ushered the concept of emotion into the modern era. Independently of one another, James and the Danish physiologist Carl Lange published theories of the emotion that both supported the idea that emotions ultimately stem from a physical movement or change.[12] James' somewhat challenging example of this idea argued that we are afraid *because* we run from the danger of an attacking bear, rather than the mental idea of fear being the incentive for movement. Contrary to current psychological concepts of how perception triggers emotional responses *before* physical movement takes place, it is movement itself that in James' thesis succeeds the initial perception of the exciting fact or object, thus preceding the emotional response. The thesis seems to imply that the bodily commotion is to be equated with the emotion. However, James soon revised his original statement to say that 'the bodily changes follow directly the perception of the exciting fact, and that our feeling of the same changes as they occur IS the emotion'.[13] Evidently, emotions became identical not with bodily changes or physical commotion, but with the *feeling* of such changes, with every emotion made up of two different elements: the bodily events and the feeling of them. James refrained from defining whether the effects that constitute the emotions are physical or non-physical in nature. Such a decision would have been metaphysical; rather than making a contribution to ontology, James' main goal was to modernise psychology, to bring to the study of emotional consciousness a set of principles from which all human emotions could be deducible.[14]

Although his revised thesis seemed to propose a dualistic theory, where the emotion is a consciousness of the bodily changes stirred by an initial excitement, James himself disowned materialistic as well as dualistic theories, conferring high status on sensations and rejecting separate faculties for the two elements of the emotion. In James' neutral monism, sensations make up the substance of both the mind and body. In spite of the fact that it was never his intention to give an answer to what the nature of consciousness is, his visceral theory has had an impact on subsequent philosophical discourse on consciousness and emotional life. I want to propose that James' ideas on movement-change cross paths with an emergent movement in philosophy referred to as an *object-oriented ontology* (hereinafter 'OOO') or *speculative realism*, known for distributing agency to what has often been disregarded as passive: the life of matter, animals and things. Furthermore, I want to connect James' theory of the emotion with a realism *without* materialism (that we know of). To elaborate on what such a philosophy would entail and what possible effects it would have on pain as lived and shared experience, I will turn to contemporary philosopher Graham Harman.

Object-oriented relations

Graham Harman belongs to a group of contemporary thinkers who are taking a renewed interest in the life of objects. Their speculative realism and materialism has attracted attention, not least from the art scene. The interest taken in OOO by the art world is in part due to the prominent role that aesthetics play in ontology, which is especially apparent in Harman's programme. It should be acknowledged that the participants of the new turn deny any connection with *naïve realism*: the idea that our sensory experience grants us a direct access to reality. The speculative is indicative of the attempt to challenge post-Kantian philosophy, which preserves the Kantian postulate that nature cannot exist (in itself) without the presence of human consciousness.[15]

In addition to opposing a philosophy of human (privileged) access, Harman claims that change cannot be explained using theories of a flat ontology of the sort that has dominated continental philosophy in its explanation of the nature of being. In a flat or flow ontology, things are only real or true as long as they enter into relationships with other things, which according to Harman implies that everything in the world is simply and fundamentally how it appears: 'Everything is language or relations or events; the thing is its action, is its effects, there is no thing hiding underneath those features.'[16] If everything relies on connectedness of the flat ontological kind, all things are effectively fixed in place, essentially defined by their relations to other things. To enable movement and change, there must be some instability in the world. Consequently, for change to happen, an asymmetrical relationship between objects must be considered. Harman's definition of objects is 'unified entities with specific qualities that are autonomous from us and from each other'.[17] An object is always independent of any objectifying or intentional act that is directed towards it. Hence, it cannot be exhausted in its qualities; there is always more to it, an inaccessible excess lurking underneath the surface. An object is never equivalent to any idea we may have of it. In fact, it is not equivalent to its own self-conception. Objects include relations, as long as they are inexhaustible and exceed any interpretation we may have of them. This implies that feelings of pain are appointed as objects since they tick all the right boxes of such entities: they are actual occurrences and they can be analysed endlessly without ever being depleted of content. The theory of how things in the world are connected is, in Harman's OOO, situated within a phenomenological life-world, but it proposes an extension beyond a still too human-centred philosophy of

Heideggerian metaphysics that has human consciousness (*Dasein*) as the principal form of intentionality. By contrast, Harman's philosophy centres on the interaction between objects and things in a world where intentionality is dispersed throughout reality.

The world as we know it is thus chopped up in vacuum-sealed pieces, but there is a world inside of vacuums too, which is the root of Harman's philosophy of relations: *vicarious causation*. To describe how the world of isolated objects relate to one another and how they create new connections without being fully disclosed or visible to other objects, Harman sketches a quadruple philosophy. The system delineates the object as composed of four poles, connected to one another through continual tension. First, there is the real object, let's say a dog, then there is the sensual object, which is the dog as caricatured or corrupted by the relation. On top of this, there is a further split in both domains. In other words, there is a division between the unity of the real dog and its real traits; qualities irrespective of our experience of the dog, and between the sensual dog and its manifold qualities; its 'public' attributes (the sensual qualities of the dog vary according to who or what is encountering it).[18] Since real objects tend to withdraw from things in the world and from a consciousness, insofar as there is one, all contact in the world must occur indirectly. Consequently, relations are always realised vicariously, mediated through a sensual object that acts as a sort of deputy. Thus, sensual objects are the glue or ether of the world, encrusting the surface of the real objects. Instead of being immersed in a world of phenomena, we seem to be bathing in a sparkling, all-too-present world of sensual objects. When two objects interact, a sensual caricature temporarily takes shape, preventing real objects from coming into direct contact with one another. According to this model, the needle of the syringe does not come into contact with the real skin; rather, it meets a soft, permeable surface.[19]

The only way for real objects to touch without actually touching is through the *allure*. Allure is a special kind of agency which connects pieces of reality with one another, generating new relations between objects by reconfiguration; dispersing, coupling and uncoupling the sensual qualities and comportments of the sensual object so as to separate or distance the sensual object from its manifold qualities. In this process, an increased tension between the real unity of an object and its sensual qualities is taking place, allowing for the sensual field to somehow become 'invaded' by the real object. The sensual object in its turn becomes 'animated by allusion to a deeper power lying beyond: a real object'.[20] The operation of evoking the allusion to the real object always falls to the share of the intending or perceiving object, which is linked to the real object only vicariously and will be 'brushing its surface in such a manner as to bring its inner life [of the real object] into play'.[21]

This might sound abstract, but such modifying processes occur frequently and include, for example, humour and metaphor in humans.[22] In relation to pain, a famous example of dispersing and reconfiguring sensual qualities in this way is Nietzsche's dog metaphor, in which he compares his pain to a dog: 'I have given my pain a name; I call it "dog".'[23] The metaphor moves the object of Nietzsche's pain to capture dog-like qualities: obtrusively barking, constantly in demand of attention and, when looking closer, faithful, entertaining and clever.[24] It should also be noted that these are all models and methods used in the making of contemporary art, in order to exaggerate, fuse or displace a particular part of sensual reality, creating new relations and connections, new resonances and understandings.

To recapitulate the story so far: for change to happen, for new objects to come into existence, connections of real objects must be realised asymmetrically. Consistent with this operation, it is always the intending object or the intentional 'I' in the perceiving act that realises the connection. The contention is that the doctor's palpating hand does not decide the effect on the real self of the patient. The cause does not decide its effect since it has but an indirect contact with the real object that it affects. I, the patient, am the one realising the effect of that touch. In the same manner, the doctor's intending hand reaching for my pained body always grasps a sensual facade. This does not mean that an emotional reaction or the outcome of a treatment cannot be calculated and estimated in advance. We are indeed object-obsessed beings and are socialised accordingly. However, the effect is nevertheless always realised by the real object exposed to the cause. Thus, an object cannot be reduced to the effect it has on me or on other things.

When the chiropractor's hand presses against my sore shoulder, adjusting it, there is not one relationship carried out, but two separate relations. As in most of our experiences, pain makes social ingress single-handedly. At first glance, it seems we can never share a real object of pain through our surface relations. However, *sincere* exchanges of pain are constantly taking place, but the 'real' here is contingent on or incidental to how we establish the sincere relation to somebody's object of pain. In 'ordinary' perception, we superficially and disinterestedly move around the exterior of other objects. Compared to such dull perceptual experience, allure is an intermittent agency that allows us to be pressed against the essence of another real object, feeling it out, so to speak. Allure implements rifts, leaps and shifts in reality. Not surprisingly, in human life allure is associated with strong emotions.[25]

Pain surfaces

To further explore the effects of allure within object-oriented interactions, we will first catch up with William James' struggle with emotional reality.[26] James rejected the allegations that his theory was materialistic on the grounds of the intricacy of emotions. The emotions are, for him, contingent on sensational processes whose nature also remains a mystery. James put embodied consciousness on a par with sensations, emotions being 'inwardly what they are, whatever physiological ground of their apparition'.[27] Thus, he defended sensations, which in his view had been subjected to philosophical prejudice, by saying that the emotions:

> carry their inner measure of work with them; and it is just as logical to use the present theory of the emotions for proving that sensational processes need not be vile and material, as to use their vileness and materiality as a proof that such theory cannot be true.[28]

Despite the fact that James' theory turns towards the commonsensical and apparently lacks concentration on the elusiveness of being, his thoughts on movement-change seem to intersect with the way in which objects and their relations are engineered in Harman's quadruple philosophy. The overlap emerges as a potent element. In Harman it is styled as a *feeling-thing* (more or less coinciding with the sensual object). Looking at James, internal bodily events become the content for consciousness to focus on when an emotion is experienced, emotion being the feeling or consciousness of bodily changes. Consequently, what emotional consciousness must be aware of is *what* is being felt. Emotions thus offer one way of being acquainted with the interior of our bodies, providing information about tacit knowledge, about the sub-representational, that which 'cannot be seen, heard, or handled, only felt'.[29] The stirring of bodily events gives rise not only to emotional awareness, but also constitutes a *felt object* that keeps interfering with emotional consciousness. In other words, the Jamesian felt object of the emotion is also a property of that emotion (of pain, for example) and has an affinity with Harman's system in the tension between the real object (of pain) and its sensual qualities. When an object becomes alluring, in that split in reality we get to catch a glimpse of the inner execution of the object itself.[30]

The natives of the sensual realm, that is to say, Harman's sensual objects, deploy themselves as colourful tentacles when I make contact with another object. Through them I can feel the conflict between the unitary reality and the specific traits of contiguous objects.[31] *Nota bene*, proximate, 'neighbourly' feuds are unceasing in such a large complex object as a human body. The rift that allure creates in the reality of the object, however, is not dissimilar to the ontological difference between the execution of being and image representation.[32] To observe somebody in pain will never be the same thing as carrying out the existence of that pain, no matter how sincerely I suffer with this individual. However, the unbridgeable gap in reality does not exclusively apply to my relationship with other beings or things. The instalment of chronic or acute pain in my own life seems to bring about the same absolute divide between introspection and embodied life, rendering every interpretation of my pain (as much as any other perceptual experience) a little askew.

Because of the receding nature of an experience so urgent as physical pain, we are keen for our fellow humans, professionals and those near and dear to us to attest to our suffering, in order to save us from the solitary confinement of pain. To have one's suffering acknowledged by any creature who is genuine as a witness to our situation is liberating. To be able to convey the tacit reality of pain myself can be even more satisfying. When Nietzsche introduces the image representation of a dog adjacent to his experience of his pain, as readers we are instantly struck with a 'dog-feeling': with its intrusive but also its amusing and wise qualities. The 'pain-dog' composite hits us in a single stroke. The metaphor actually makes us live a new, unexpected feeling-object.[33]

Reality check

What are the implications of an object-oriented ontology for the experiences of mediating and sharing physical pain and hurt?

Sharing pain

First, the notion of 'sharing pain' must be understood equally as the relationship one maintains with oneself as well as the one we enter into with objects and things in the outside world. In accordance with Harman's OOO, I have suggested that to obtain a deepened understanding of the representability of pain, its social and political potential, pain relations should be developed using techniques of allure/allusion. The example brought up for the purpose of this exercise is Nietzsche's juxtaposition of pain with a dog. Nietzsche cannot reach to the nethermost reality of his pain, but he is sincerely grabbing at its effects: its shapes, tones and colours. He is separating the qualities of his pain from its essence, thereby thrusting the reader into a new relationship with pain and, ideally, to new ways of approaching illness. He is here ingeniously making manifest the liberating powers of embodied thinking.

Critique vs. fascination

Harman's OOO does not make a big critical statement compared to some of the more outspokenly radical movements that pre-date speculative realism. Concurrently, the movement has met with criticism. Accused of reversion to modernistic and romantic ideals by bringing individual substances back, one would think objects are desiring and reifying humans, and not the other way around.[34] What is to be said about such highly motivated criticism?

To focus on the alluring aspects of pain may very well contribute to creating a spectacle. However, if objects cannot be exhausted by their relations, it means we cannot be reduced to the relationships we enter into or to any feature that we are told defines us.[35] Unremitting change is something one would want to take comfort in. It implies that we are not our diagnosis, or the lack of a diagnosis. We will always transcend our contexts, however confining they may seem. On the other hand, there will always be a mismatch between self-interpretation and reality, but underneath our experiences there will still be a real self, rich enough to maintain many possible interpretations. Ultimately, we contain the boundlessness of reality within ourselves.

Harman supports his shift from a philosophy of critique to a 'philosophy of fascination' by crossing paths with William James for the first time, paraphrasing him by saying: 'What we really need are not more critical readers, but more vulnerable ones, readers so hungry for the unexpected that they can "recognise a good [idea] when they see [it]".'[36] With these words, OOO seems to be offering a line of flight from fixed positions. There is a window left ajar for us to open up towards the fascinating realities of the comical, cute, mystical and the downright bizarre.

Aesthetic effects

The turn towards objects in OOO is also concerned with a redistribution of embodied capacities, to include the agency of inanimate things. With allure permeating not only animals (humans included), but also all reality in germinal form, in Harman's programme aesthetics is designated first philosophy. All objects aestheticise themselves. When fire burns cotton, fire is behaving in a certain way towards cotton that acts as a vicarious cause for fire's consuming trait. When lingering too long over the flame of a candle, my wrist acts as a vicarious cause for fire's burning trait. The aesthetic effect of an object is dependent on the vicarious cause, on the adjacent object with which it is entering into relation. In the case of an artwork, we would probably refer to its alluring, enchanting and aestheticising agency as if it had a certain *style*. Harman also clearly makes the connection between what we call style in art, which is nothing other than 'a specific mode or method of de-creating images and re-creating them as feeling-things'.[37] The animating style of an artwork, Harman remarks, makes it analogous to the human body. As we have seen, in object-oriented interactions we cannot grasp either the artwork or the body in itself, we can only grasp its style. But we can allude to its content in an inconceivable number of ways.

In conclusion, my work with *Documenting Bodies* was an attempt to bring together medical objects and pain realities so as to allow for formal as well as grammatical qualities inherent in the content to transform the social realities conveyed in the work. It features no undistorted feelings of 'real pain'. What there is, though, is the remodelling of the different elements and their qualities into hybrid structures, which hopefully will cater to the bodies of the audience, who will live new feeling-objects.

Object-oriented ontology encourages us to exercise a non-anthropocentric (that is, extra-linguistic), aesthetic attitude towards pain. Before the real object surfaces in the sensual sphere, invoking new, shimmering qualities and resonances in our body sounding boards, it is implied that there is always already an agency such as allure at play before a human mind makes 'something out of nothing', where 'nothing' here refers to the idea of a pure sentience of pain.[38] To be gazed back at from the unseen is to accept that 'nothing' is already always some*thing*.

Notes

1. It should be noted that both the art project and this text disown the idea that first-person accounts, and explorations thereof, would have some kind of privileged access to experience. Rather, they are both attempts at problemising the relationships between first- and third-person perspectives. At the same time, they challenge the notion that supports the primacy of human mind and language in the communication of pain by opening up to the idea that an 'inner point of view' of being does not necessarily entail a human consciousness. See Franciso J. Verela and Jonathan Shear,

'First-Person Methodologies: What, Why, How?', *Journal of Consciousness Studies*, 6(2–3) (1999): 1–14.

2. Martha Joanna Zarzycka, 'Body as Crises, Representations of Pain in Visual Arts' (PhD thesis, Utrecht University, 2007), 5–6. See also David Biro, Chapter 4, this volume.

3. See Joanna Bourke, Chapter 5, this volume and Wilfried Witte, Chapter 6, this volume.

4. Patricia Ticineto Clough (ed.), *The Affective Turn: Theorizing The Social* (Durham, NC: Duke University Press, 2007), 2.

5. Gilles Deleuze and Félix Guattari, *What is Philosophy?* (trans. Hugh Tomlinson and Graham Burchell) (New York: Columbia University Press, 1994), 164.

6. Arthur W. Frank, *The Wounded Storyteller: Body, Illness, and Ethics* (University of Chicago Press, 1995), 5–7.

7. *Ibid.*, 5.

8. See Whitney Wood, Chapter 11, this volume.

9. Marianne Torp, 'Reality Re-checked: Dokuficiering, re-enactment og objektets rekontekstualisering i nyere kunst', in Sven Bjerkhof (ed.), *Reality Check* (Copenhagen: Statens Museum for Kunst, 2008), 8.

10. *Ibid.*, 8.

11. Jane Bennett, *Empathic Vision: Affect, Trauma, and Contemporary Art* (Stanford University Press, 2005), 10.

12. Gerald E. Meyer, *William James: His Life and Thought* (New Haven: Yale University Press, 1986), 220.

13. William James, *The Principles of Psychology* (New York: Henry Holt & Co., 1905 [1890]), vol. 2, 449.

14. Meyer, *William James*, 216.

15. Bryant Levi, Nick Srnicek and Graham Harman, 'Towards a Speculative Philosophy', in Bryant Levi, Nick Srnicek and Graham Harman (eds), *The Speculative Turn: Continental Materialism and Realism* (Melbourne: re.press, 2011), 4.

16. Graham Harman, 'Critique is *Still* Out of Steam', Lecture given at *The Faculty Collegium for Critique and the Valand Seminar for Advanced Art Theory: New Focuses in Artistic Practice and Critical Writing within Contemporary Art* (12 October 2012), Valand Academy, Gothenburg.

17. Graham Harman, 'On the Undermining of Objects: Grant, Bruno, and Radical Philosophy', in Levi, Srnicek and Harman (eds), *The Speculative Turn*, 22.

18. Graham Harman, *Guerrilla Metaphysics: Phenomenology and the Carpentry of Things* (Chicago: Open Court, 2005), 77, 153.

19. Fredrik Österblom, 'Graham Harman och den objekt-orienterade ontologin', *Paletten*, 289 (2012): 26–9 at 28.

20. Graham Harman, 'On Vicarious Causation', *Collapse II: Speculative Realism* (2007): 187–221 at 220.

21. *Ibid.*, 220.

22. Harman, *Guerrilla Metaphysics*, 216.

23. Friedrich Nietzsche, *Den glada vetenskapen* (trans. Carl-Henning Wijkmark) (Gothenburg: Korpen, 2011), 207 (my translation).

24. *Ibid.*, 207.

25. Harman, *Guerrilla Metaphysics*, 213–14.

26. For the reception of James' theory, I am using Gerald E. Meyer's analyses throughout the text.
27. James, *The Principles of Psychology*, 453.
28. *Ibid.*, 453.
29. Meyer, *William James*, 236.
30. Harman, *Guerrilla Metaphysics*, 161.
31. *Ibid.*, 174.
32. *Ibid.*, 110.
33. *Ibid.*, 109. We live a new feeling-object, but not the real object itself, since a real object can never be lived by any other thing. The use of metaphorical language to describe the experience of pain is a precarious act, to which Susan Sontag paid detailed attention in her 1978 acclaimed study *Illness as Metaphor* (New York: Farrar, Straus & Giroux). Quite often one is liable to cement rather than to break up preconceived notions about pain, to give way to generalised metaphors that reinforce a mind-body split by distancing internal bodily events from conscious reflection (the use of warfare metaphors in pain narratives, for example). The anthropomorphism that Nietzsche attributes to the real object of his pain succeeds in eliminating such static generalisations of pain by swiftly modifying the terrain of physical suffering.
34. Sinziana Ravini, 'Ontologiska slagfält vid modernitetens gräns', *Paletten*, 289 (2012): 30–9 at 39.
35. See also Österblom, 'Graham Harman', 29.
36. Harman, *Guerrilla Metaphysics*, 239. The full quote in James is: 'The best claim a college education can possibly make on your respect, the best thing it can aspire to accomplish for you, is this; that it should help you to know a good man when you see him' (William James, 'The Social Value of the College Bred', in *Memories and Studies* (Rockville, MD: Arc Manor, 2008), 124).
37. Harman, *Guerrilla Metaphysics*, 110.
38. Elaine Scarry, *The Body in Pain: The Making and Unmaking of the World* (Oxford University Press, 1985), 280.

Select Bibliography

The diversity of material in a volume such as this, with regards to geographical, historical and disciplinary coverage, would make a comprehensive bibliography both unwieldy and unhelpful. I refer the reader to the notes appended to each chapter for references specific to each thematic and disciplinary approach. There are, however, a number of works around which all the disciplines represented here seem to pivot.

For comprehensive histories of pain, with coverage from antiquity to the present, see David B. Morris, *The Culture of Pain* (Berkeley: University of California Press, 1991), Roselyne Rey, *The History of Pain* (Cambridge, MA: Harvard University Press, 1998) and Thomas Dormandy, *The Worst of Evils: The Fight Against Pain* (New Haven: Yale University Press, 2006). Another *longue durée* perspective, but with themes much in accord with this book, is the recent volume *Knowledge and Pain* (Amsterdam: Rodopi, 2012), edited by Esther Cohen, Leona Toker, Manuela Consonni and Otniel Dror. For a treatment on the long history of psychological healing and its relation to pain, see S.W. Jackson, *Care of the Psyche: A History of Psychological Healing* (New Haven: Yale University Press, 1999). For a more modern focus and a cultural approach, Javier Moscoso has defined the path in *Pain: A Cultural History* (Basingstoke: Palgrave Macmillan, 2012). Joanna Bourke already contributed significantly to the modern cultural history of pain in *What it Means to Be Human: Reflections from 1791 to the Present* (London: Virago, 2011), but this will be greatly augmented by her *The Story of Pain: From Prayer to Painkillers* (Oxford University Press, 2014).

More specific histories, which nevertheless utilise approaches that have wider implications, include Lucy Bending, *The Representation of Bodily Pain in Nineteenth-Century English Culture* (Oxford University Press, 2000) and Andrew Hodgkiss, *From Lesion to Metaphor: Chronic Pain in British, French and German Medical Writings, 1800–1914* (Amsterdam: Rodopi, 2000). Scholars associated with the Birkbeck Pain Project in London contributed their 'Perspectives on Pain' to a wide-ranging special issue of *19: Interdisciplinary Studies in the Long Nineteenth Century*, 15 (2012), edited by Joanna Bourke, Louise Hide and Carmen Mangion.

Though it has no historical input, the recent volume edited by Lisa Folkmarson Käll, *Dimensions of Pain: Humanities and Social Science Perspectives* (London: Routledge, 2013) explores 'pain studies' from a variety of humanities and social-science perspectives that intersect with the focus of this book. The archetype of such studies, of course, is Elaine Scarry, *The Body in Pain* (Oxford University Press, 1987). Though now subject to revisionism, it is still the yardstick by which other studies on pain measure themselves.

Medical and scientific interventions in the study of pain and emotion are vital to the success of the interdisciplinary perspective. The most significant introduction to pain is perhaps still R. Melzack and P.D. Wall, *The Challenge of Pain* (London: Penguin, 1996). Those whose focus on emotion most usefully translates to the study of pain are William Reddy, *The Navigation of Feeling: A Framework for the History of Emotions* (Cambridge University Press, 2001) and Jerome Kagan, *What is Emotion? History, Measures, and Meanings* (New Haven: Yale University Press, 2007). Combining both, Nikola Grahek, *Feeling Pain and Being in Pain* (2nd edn, Cambridge, MA: MIT Press, 2007) is essential reading.

Index